London Mathematical Society Student Texts. 9

Automorphisms of Surfaces after Nielsen and Thurston

Andrew J. Casson, University of California, Berkeley
and
Steven A. Bleiler, University of British Columbia

CAMBRIDGE
UNIVERSITY PRESS

CAMBRIDGE UNIVERSITY PRESS
Cambridge, New York, Melbourne, Madrid, Cape Town, Singapore,
São Paulo, Delhi, Dubai, Tokyo, Mexico City

Cambridge University Press
The Edinburgh Building, Cambridge CB2 8RU, UK

Published in the United States of America by
Cambridge University Press, New York

www.cambridge.org
Information on this title: www.cambridge.org/9780521349857

First published 1988
Reprinted 1993

A catalogue record for this publication is available from the British Library

Library of Congress Cataloguing in Publication Data

Casson, A. (Andrew)
Automorphisms of Surface after Nyfisen & Thurston
(A. Casson & S. Bleiler)
p. cm.- (London Mathematical Society student texts; 9).
Bibliography: p. Includes index
1. Surfaces. 2. Automorphism/. 3. Geometry, Hyperbolic.
I. Bleiler, S. II. Title III. Serie
QA685.C341988 516-dco 87–30938

ISBN 978-0-521-34203-2 Hardback
ISBN 978-0-521-34985-7 Paperback

LONDON MATHEMATICAL SOCIETY STUDENT TEXTS

1 Introduction to combinators and λ-calculus, J.R. HINDLEY & J.P. SELDIN
2 Building models by games, WILFRID HODGES
3 Local fields, J.W.S. CASSELS
4 An introduction to twistor theory, S.A. HUGGETT & K.P. TOD
5 Introduction to general relativity, L. HUGHSTON & K.P. TOD
6 Lectures on stochastic analysis: diffusion theory, DANIEL W. STROOCK
7 Dynamical systems and the theory of evolution, J. HOFBAUER & K. SIGMUND
8 Summing and nuclear norms in Banach space theory, G.J.O. JAMESON
9 Automorphisms of surfaces after Nielsen and Thurston, A.CASSON & S. BLEILER
10 Non-standard analysis and its applications, N.CUTLAND (ed)
11 The geometry of spacetime, G. NABER
12 Undergraduate algebraic geometry, MILES REID

Preface

During the Fall of 1982 and Spring of 1983, Andrew Casson gave a course in surface automorphisms at the University of Texas. This book began as my lecture notes from that course. Andrew and I have revised these notes into the present form. With this book we hope not only to give students an introduction to hyperbolic geometry and its applications to low-dimensional topology, but also provide a concise exposition of the Nielsen-Thurston theory of surface automorphisms. The reader should have an understanding of the basic lecture courses in topology and linear algebra. We begin by recalling the structure of the self-homeomorphisms of the torus.

CONTENTS

§0. Introduction

Our emphasis is on closed orientable surfaces, usually denoted by F, with the *genus* of F, the number of torus summands, denoted by g. An *automorphism* of a surface F we take to be a homeomorphism $h: F \to F$, usually orientation-preserving. The Nielsen–Thurston theory generalizes the well-known classification of toral automorphisms to the automorphisms of an arbitrary orientable surface. We begin by briefly recalling this classification.

Regard the torus as the quotient of the Euclidean plane $\mathbb{R}^2$ by the integer lattice $\mathbb{Z}^2$, endowed with a fixed orientation. Hence $\pi_1(T^2) \cong \mathbb{Z}\oplus\mathbb{Z}$. The homeomorphisms of T^2 correspond to the elements of the general linear group $GL_2(\mathbb{Z})$ as any element α in $GL_2(\mathbb{Z})$ maps $\mathbb{Z}^2$ to itself and so induces a continuous map h_α: $T^2 \to T^2$. The homeomorphism h_α has inverse $h_{\alpha}-1$ and $(h_\alpha)^*: \pi_1(T^2) \to \pi_1(T^2)$ has matrix α. The map h_α preserves orientation if and only if $\det(\alpha) = 1$. i.e. α is an element of the special linear group $SL_2(\mathbb{Z})$.

If α in $SL_2(\mathbb{Z})$ is represented by the 2x2 matrix $\begin{bmatrix} p & q \\ r & s \end{bmatrix}$, the characteristic polynomial of α, $t^2-(p+s)t+(ps-rq)$, can be written as $t^2-(\mathrm{trace}(\alpha))t+1$. The eigenvalues λ, λ^{-1} of α are either

1) complex, i.e. $\mathrm{trace}(\alpha) = 0, 1,$ or -1

or 2) both ± 1, i.e. $\mathrm{trace}(\alpha) = \pm 2$

or 3) distinct reals, i.e. $|\mathrm{trace}(\alpha)| > 2$.

Consider each of these cases in turn.

In case 1) an easy exercise in the Cayley–Hamilton theorem shows that α is of finite order and $(h_\alpha)^{12} = 1$. The map h_α is said to be *periodic*.

In case 2) α has an integral eigenvector which projects to an essential simple closed curve C under the quotient map $\mathbb{R}^2 \to T^2$. The curve C is invariant under the map h_α, but possibly with reversed string orientation. In this case the map h_α is a power of a *Dehn twist* in C. More precisely, the curve C has a regular neighborhood A homeomorphic to an annulus; which we consider parameterized as $\{[r,\theta] \mid 1 \leq r \leq 2\}$. The *Dehn twist* in C is defined to be the homeomorphism given by the identity off A and by the map $[r,\theta] \to [r,\theta+2\pi r]$ on A, see Figure 0.1.

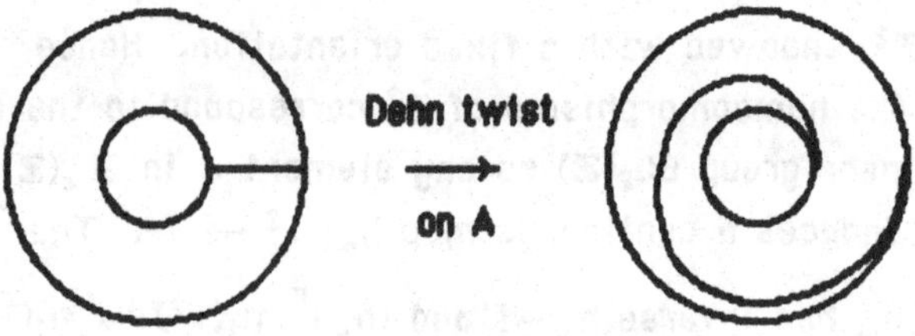

Figure 0.1

In this case the map h_α is said to be *reducible*.

For case 3) suppose that $|\lambda| > 1 > |\lambda^{-1}|$ and that x and x' are the corresponding real eigenvectors. The map h_α has infinite order and leaves no simple closed curve invariant. Translating the vectors x, x' yields vector fields $\mathcal{F}, \mathcal{F}'$ which are carried by h_α to vector fields $\lambda\mathcal{F}$ and $\lambda^{-1}\mathcal{F}'$ respectively. That is, h_α is a linear homeomorphism which stretches by a factor λ in one direction and shrinks by the same factor in a complementary direction. When this occurs h_α is called *Anosov*.

§1. The Hyperbolic Plane $\mathbb{H}^2$

This chapter contains all the necessary background material on hyperbolic plane geometry. Much of the work generalizes to higher dimensions. We use the Poincare disk model which identifies the hyperbolic plane with the interior of the unit disk D^2 in the Euclidean plane $\mathbb{R}^2$. The boundary of D^2 is the *circle at infinity* S^1_∞; notice that $\mathbb{H}^2 \cap S^1_\infty$ is empty.

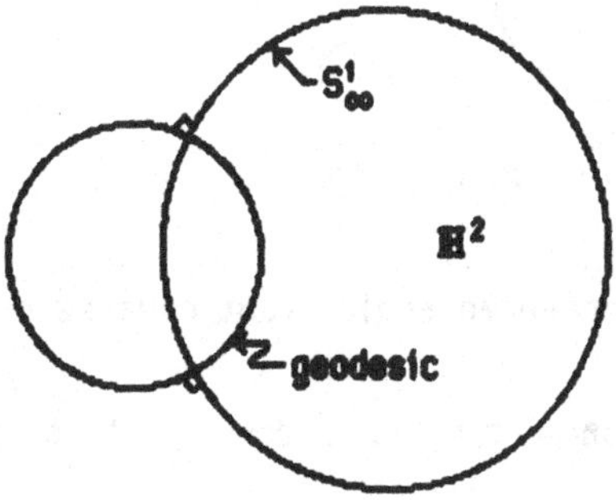

Figure 1.1

It is convenient to regard $\mathbb{R}^2$ as embedded in its one-point compactification $\mathbb{R}^2 \cup \{\infty\}$, so that Euclidean lines can be regarded as circles through ∞. A *geodesic* (or straight line) in $\mathbb{H}^2$ is $C \cap \mathbb{H}^2$, where C is a circle in $\mathbb{R}^2 \cup \{\infty\}$ meeting S^1_∞ orthogonally. There is a unique geodesic joining any two points in $\mathbb{H}^2$. The *angle* between two intersecting geodesics is the Euclidean angle between their defining circles.

Next, the operation of *reflection* in a geodesic of $\mathbb{H}^2$ will be defined in terms of Euclidean inversion. If C is a circle in $\mathbb{R}^2$ with center O and radius r, *inversion* in C carries a point $P \in \mathbb{R}^2-\{O\}$ to the unique point P′ on the ray OP such that $OP \cdot OP' = r^2$, and interchanges O with ∞. If C is a line in $\mathbb{R}^2$, inversion in C is just Euclidean reflection in C. In either case, inversion in C defines an involution of $\mathbb{R}^2 \cup \{\infty\}$.

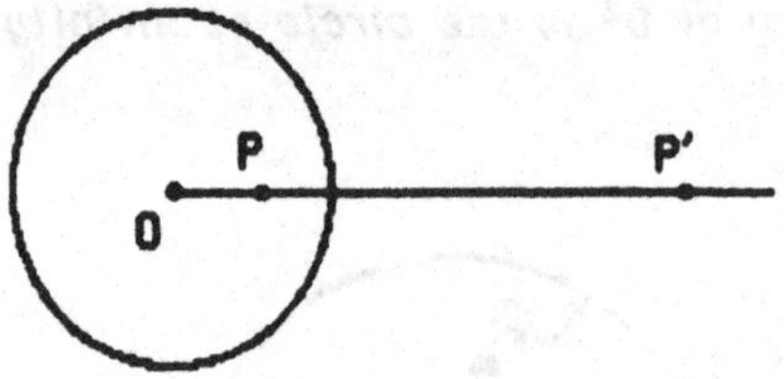

Figure 1.2

Lemma 1.1: Inversions preserve angles (but reverse orientation).

Lemma 1.2: Inversions carry circles in $\mathbb{R}^2 \cup \{\infty\}$ to circles in $\mathbb{R}^2 \cup \{\infty\}$.

The proofs are (well-known) exercises in Euclidean geometry.

If $C \cap \mathbb{H}^2$ is a geodesic in $\mathbb{H}^2$ then inversion in C induces an involution of $\mathbb{H}^2$, called *reflection* in $C \cap \mathbb{H}^2$. An *isometry* of $\mathbb{H}^2$ is defined to be a product of reflections. By 1.1, isometries preserve angles.

Lemma 1.3: The group of isometries acts transitively on $\mathbb{H}^2$, and the stabilizer of any point in $\mathbb{H}^2$ is isomorphic to O(2).

Proof: Let O be the center of D, A any point of $\mathbb{H}^2$. Then A can be carried to O by a single reflection in a geodesic $C \cap \mathbb{H}^2$, C having center on the ray OA. So any two points of $\mathbb{H}^2$ are "connected" by an isometry that is the product of at most two reflections.

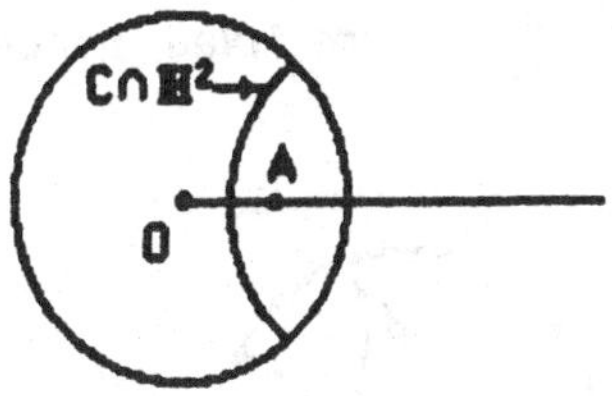

Figure 1.3

In light of this it follows that the stabilizers of any two points in $\mathbb{H}^2$ are isomorphic via conjugation. Thus we only need determine Stab(O). This group contains the reflections in the lines through O. A rotation about O can be expressed as the product of two reflections and these generate O(2).

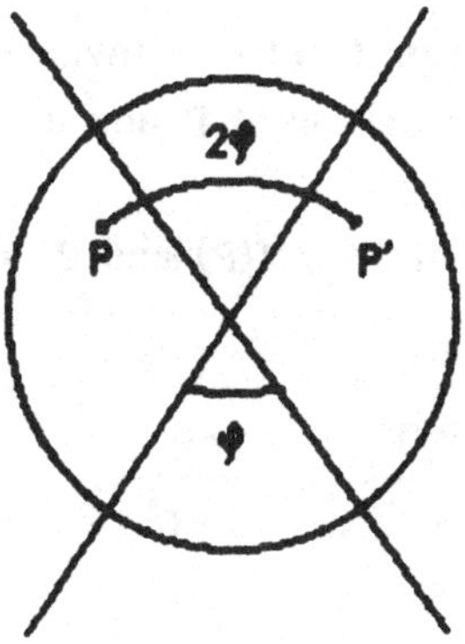

Figure 1.4

To see that O(2) is the entire stabilizer of 0, notice that isometries extend in a unique manner to S^1_∞. So it is enough to show that any isometry extending to the identity on S^1_∞ is actually the identity. But for any point P in $\mathbb{H}^2$ which is the intersection of geodesics γ,γ', the "ends" of γ,γ' are fixed, hence γ,γ' are fixed, and hence P.

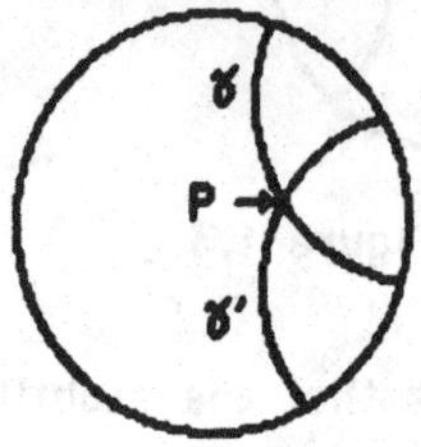

Figure 1.5

Lemma 1.4: Isometries leave $ds/(1-r^2)$ invariant where r denotes Euclidean distance from the center of D and ds=d(Euclidean distance).

Proof: We must show that if P' = f(P) and Q' = f(Q), where f is an isometry of $\mathbb{H}^2$, then

$$\frac{P'Q'}{1-r'^2} \approx \frac{PQ}{1-r^2}.$$

It follows from Lemma 1.3 that it is enough to check this when P is the center of the Poincare disk and f is a reflection in $C \cap \mathbb{H}^2$ (where C is a Euclidean circle with center O and radius k).

By properties of inversion, $\dfrac{P'Q'}{PQ} \approx \dfrac{OP'}{OP}$ when Q is near P.

From Figure 1.6 $\dfrac{OP'}{OP} = \dfrac{k^2}{(OP)^2} = 1-r'^2.$

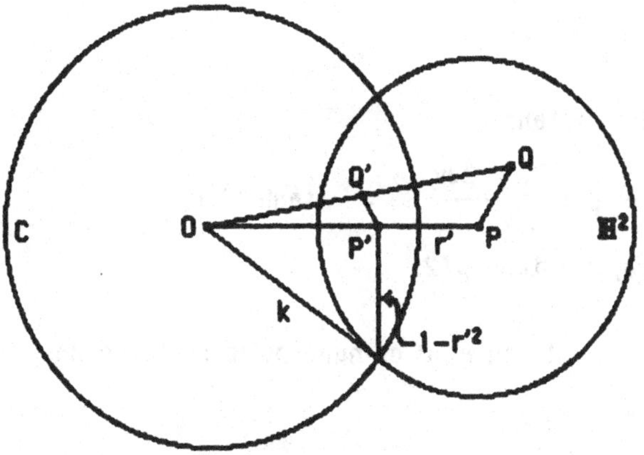

Figure 1.6

Hence $\dfrac{P'Q'}{1-r'^2} \approx \dfrac{PQ}{1-0^2}$.

Exercise: Show that geodesics with respect to this metric really are geodesics.

Remark: It suffices to prove that lines through the center of the Poincare disk minimize hyperbolic arc length.

Definition: The *hyperbolic metric* on $\mathbb{H}^2$ is given by $2ds/(1-r^2)$.

Convention: Greek letters denote hyperbolic distance, Roman letters denote Euclidean distance.

Examples:

1. Hyperbolic distance

$$OP = \rho = \int_0^r \frac{2dx}{1-x^2} = 2\tanh^{-1} r$$

Therefore, $r = \tanh \rho/2$.

2. The circle centered at O of hyperbolic radius ρ has circumference

$$2\pi \frac{2r}{1-r^2} = \frac{4\pi r}{1-r^2} = 2\pi \tanh \rho/2 \cosh^2 \rho/2 = 2\pi \sinh \rho.$$

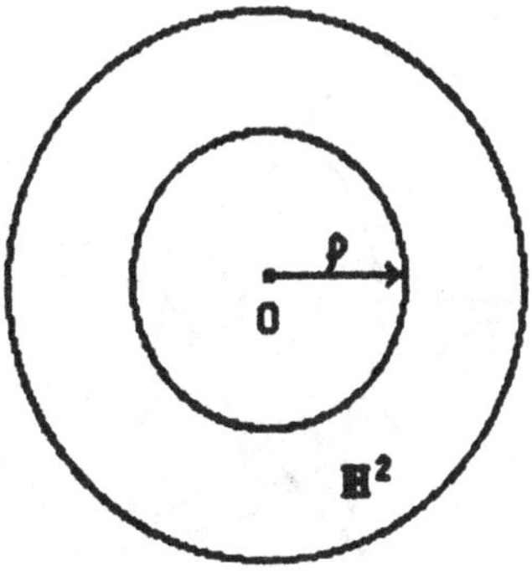

Figure 1.7

3. The circle of the previous example has area

$$\int_0^r \frac{4\pi x}{(1-x^2)^2} 2dx = 4\pi\left(\frac{1}{1-r^2} - 1\right) = 2\left(\frac{2r^2}{1-r^2}\right)$$

In terms of hyperbolic distance, this is

$$\int_0^\rho 2\pi \sinh\tau \, d\tau = 2\pi(\cosh\rho - 1).$$

Theorem 1.5: (Gauss–Bonnet) A geodesic triangle in $\mathbb{H}^2$ with angles α, β, γ has area $\pi - (\alpha+\beta+\gamma)$.

Proof: Without loss of generality the α vertex is at the center of the Poincare disk, and as we can subdivide our triangles into right triangles we may assume $\gamma = \pi/2$. First note

$$k\sin\beta \, d\theta = r d\alpha \Rightarrow \frac{d\theta}{d\alpha} = \frac{r}{k\sin\beta} = \frac{2r}{1/r - r} = \frac{2r^2}{1-r^2}.$$

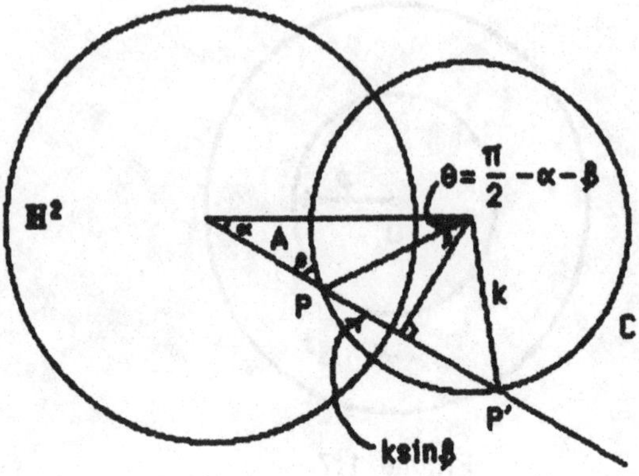

Figure 1.8

Further note: $\dfrac{dA}{d\alpha} = \dfrac{2r^2}{1-r^2}$.

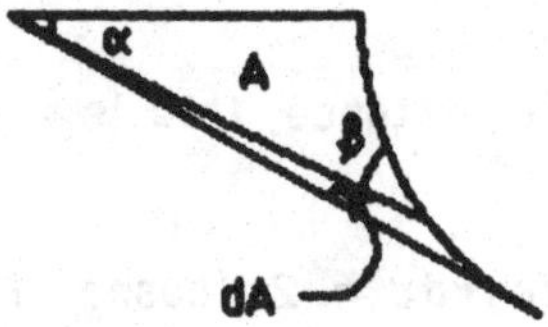

Figure 1.9

Thus $A = \theta + \text{constant} = \pi/2 - \alpha - \beta + \text{constant}$. When $\alpha=0$ then $\beta=0$, so the constant is zero.

Corollary 1.5.1: An n-gon with angles $\alpha_1, \ldots, \alpha_n$ has area

$$(n-2)\pi-(\alpha_1 + \ldots + \alpha_n).$$

Corollary 1.5.2: If the vertices of an n–gon are all on the circle at ∞, the area is $(n-2)\pi$.

Lemma1.6: Let $h:\mathbb{H}^2 \rightarrow \mathbb{H}^2$ be an orientation preserving isometry, $h \neq 1$. Then exactly one of the following occurs.

1. h has a unique fixed point in $\mathbb{H}^2$, no fixed points on S^1_∞. Call such h *elliptic*.
2. h has a unique fixed point on S^1_∞, none in $\mathbb{H}^2$. Call such h *parabolic*.
3. h has exactly two fixed points on S^1_∞, none in $\mathbb{H}^2$. Call such h *hyperbolic*.

Remark: A similiar result for $h:\mathbb{H}^n \rightarrow \mathbb{H}^n$ can be proved the same way.

Proof: By Brouwer's theorem, $h:D^2 \rightarrow D^2$ has a fixed point.

Case 1: If there is an interior fixed point, we may assume that this point is the center of D. Hence h is a rotation and we are in case 1 above.

Case 2: If h fixes two (or more) points P and Q on the circle at infinity, h leaves invariant the unique geodesic PQ. The map h restricted to this geodesic is translation through hyperbolic distance d. Call this invariant geodesic the *axis* of h.

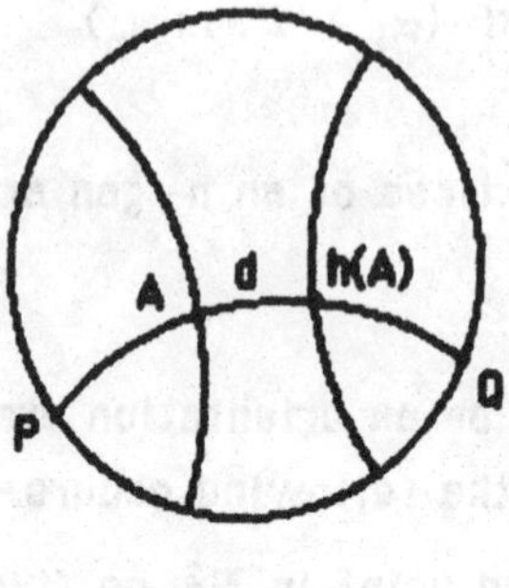

Figure 1.10

This determines h on the circle at infinity as perpendicular geodesics go to perpendicular geodesics, hence h has exactly two fixed on the circle at infinity, hence h is as in case 3 above.

All other h's fall into case 2 above.

The half-plane model of $\mathbb{H}^2$

Instead of the disc D we can apply a sequence of inversions and use the upper half-plane in $\mathbb{C}$ as our model of $\mathbb{H}^2$. In this model the circle at infinity is $\mathbb{R}\cup\{\infty\}$ and geodesics are vertical lines and circles meeting $\mathbb{R}$ in right angles. The invariant metric is ds/y where y is the imaginary part.

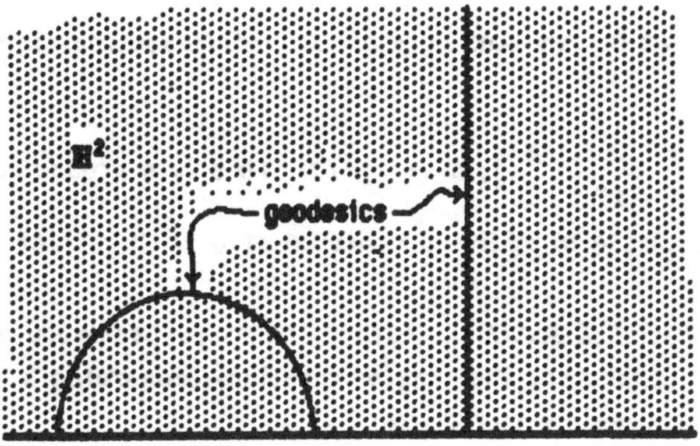

Figure 1.11

Theorem 1.7: The group of orientation preserving isometries of $\mathbb{H}^2$ is isomorphic to $PSL_2(\mathbb{R})$.

Recall: $SL_2(\mathbb{R})$ is the set of real 2 x 2 matrices with determinant 1. $PSL_2(\mathbb{R}) = SL_2(\mathbb{R})/\pm I$.

A matrix $\begin{bmatrix} a & b \\ c & d \end{bmatrix}$ in $PSL_2(\mathbb{R})$ acts on the upper half plane via

$$z \longrightarrow \frac{az+b}{cz+d} .$$

Note that this preserves the real axis.

Proof: $PSL_2(\mathbb{R})$ is generated by matrices of the following two forms.

First: $\begin{bmatrix} 1 & b \\ 0 & 1 \end{bmatrix}$, $z \longrightarrow z + b$. This is a translation parallel to the real axis and can be realized by a product of reflections in vertical geodesics.

Second: $\begin{pmatrix} 0 & a \\ -a^{-1} & 0 \end{pmatrix}$, $z \to -a^2/z$. This is a product of an inversion in a circle of radius a centered at 0 with reflection in the Y-axis.

Hence $PSL_2(\mathbb{R}) \subseteq Aut(\mathbb{H}^2)$. Follow the steps backward and it is easy to check (**Exercise**) that any orientation-preserving h can be represented by a matrix in $PSL_2(\mathbb{R})$.

Remark: An exercise in linear algebra recovers Theorem 1.6.

We wish to develop conditions for an isometry to be elliptic, hyperbolic, or parabolic. Given h, represented as a matrix in $PSL_2(\mathbb{R})$ as above, the equation for fixed points,

$$z = \frac{az+b}{cz+d},$$

implies that $cz^2+(d-a)^2-b = 0$. The discriminant of this equation is easily seen to be $(\text{trace})^2 - 4$, where trace = a+d.

Case 1: If $|\text{trace}| < 2$ then the roots are complex, so there is exactly one fixed point in $\mathbb{H}^2$. In this case, h is elliptic and represented by a rotation in the Poincare disk model.

Case 2: If $|\text{trace}| > 2$ then the roots are real, so there are exactly two fixed points on the circle at infinity and h is hyperbolic. To represent h use the upper half-plane model with $0,\infty$ as fixed points. The axis of h is the imaginary axis.

If $z \to \dfrac{az+b}{cz+d}$ fixes $0,\infty$ then $b=c=0$ and the matrix is $\begin{bmatrix} a & 0 \\ 0 & a^{-1} \end{bmatrix}$ i.e. $z \to a^2 z$; h is a *dilation.*

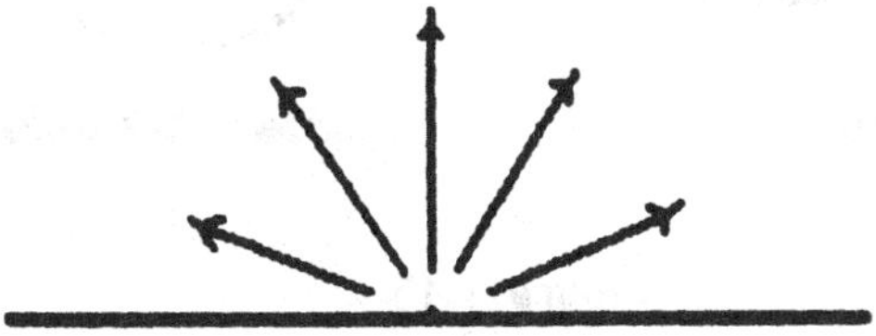

Figure 1.12

Case 3: If $|\text{trace}| = 2$ there are two equal real roots so h has a unique fixed point on the circle at infinity and is therefore parabolic. To represent h use the half-plane model with ∞ as fixed point.

If $z \to \dfrac{az+b}{cz+d}$ fixes ∞, then $c=0$ and the matrix is $\begin{bmatrix} a & b \\ 0 & 1/a \end{bmatrix}$, where $a=\pm 1$ since $a+a^{-1} = \pm 2$. Thus the matrix for the general parabolic isometry fixing ∞ is given by $\begin{bmatrix} 1 & b \\ 0 & 1 \end{bmatrix}$. But this is just $z \to z+b$, horizontal translation parallel to the real axis.

So a parabolic isometry fixing ∞ preserves horizontal levels in the half-plane model. These levels are called *horocycles.* In the Poincare disk model horocycles are Euclidean circles in D tangent to S^1_∞.

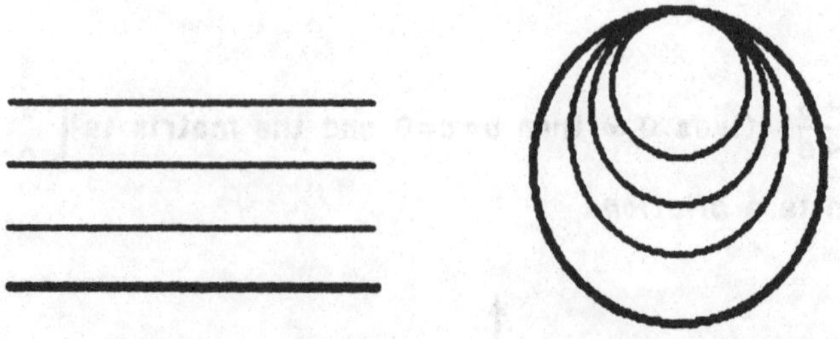

Figure 1.13

Remark: The group of orientation preserving isometries of $\mathbb{H}^3$ is isomorphic to $PSL_2(\mathbb{C})$, which as before is $SL_2(\mathbb{C})/\pm I$, where $SL_2(\mathbb{C})$ is the set of 2×2 complex matrices with determinant 1.

A matrix $\begin{bmatrix} a & b \\ c & d \end{bmatrix}$ in $PSL_2(\mathbb{C})$ acts on the "floor" plane $\mathbb{C}$ via

$$z \longrightarrow \frac{az+b}{cz+d},$$

but note that here the real axis is *not* preserved as a,b,c,d are complex. $PSL_2(\mathbb{C})$ is generated by $z \to z+b$, $z \to -a^2/z$; $a,b \in \mathbb{C}$. Notice $z \to z+b$ can be realized by two reflections in vertical hyperplanes in Euclidean space, while $z \to -a^2/z$ can be realized by inversion in a Euclidean sphere. So these extend uniquely to isometries of $\mathbb{H}^3$, hence any element in $PSL_2(\mathbb{C})$ extends. As before, the uniqueness of the extension is an intersecting geodesics argument.

§2. Hyperbolic Structures on Surfaces

In this chapter we study hyperbolic structures on surfaces. We shall be primarily concerned with closed surfaces, but we shall also consider hyperbolic structures of finite area on other surfaces.

Definition: A *hyperbolic structure* on a surface F is determined by an atlas of charts

$$\varphi_\alpha\colon U_a \to \mathbb{H}^2 \text{ such that } \varphi_\beta\varphi_\alpha^{-1}\colon(\varphi_\alpha(U_\alpha \cap U_\beta)) \to \varphi_\beta(U_\alpha \cap U_\beta)$$

is the restriction of an orientation-preserving isometry of $\mathbb{H}^2$.

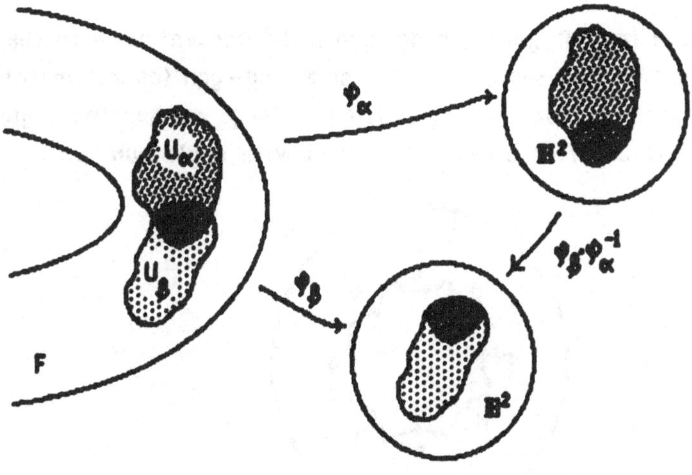

Figure 2.1

We usually require F and the atlas to be oriented.

Example: Every closed oriented surface F of genus > 1 has a hyperbolic structure. The surface F is made by identifying edges of a 4g-gon (so that all vertices are identified).

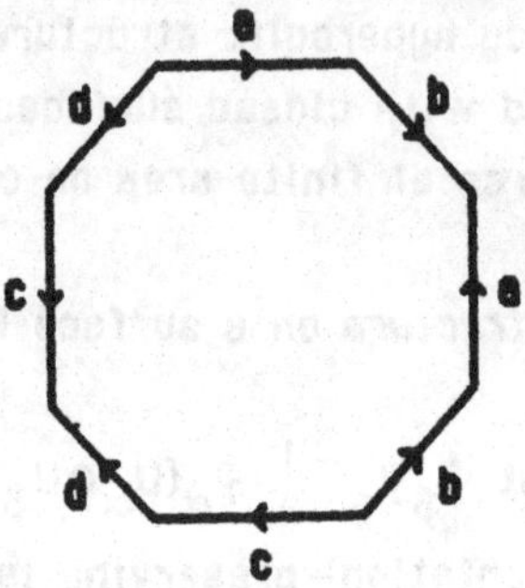

Figure 2.2

Consider a regular geodesic 4g-gon in H^2 concentric with the Poincare disk D^2. The angle sum of a small 4g-gon (approximately $(4g-2)\pi$) is greater than 2π if $g > 1$. For a large 4g-gon the angle sum is small, so we can find a regular 4g-gon with angle sum 2π.

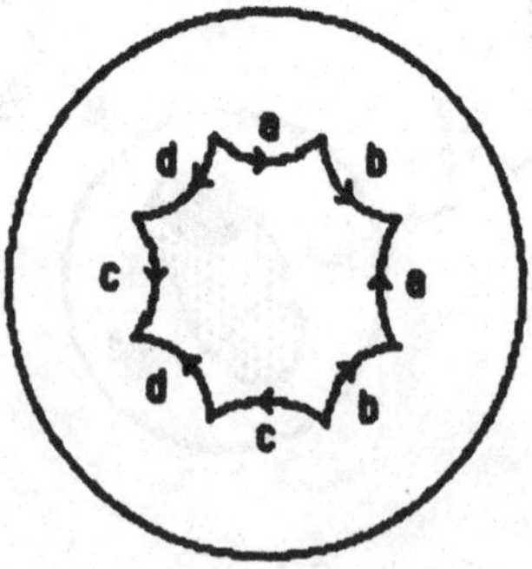

Figure 2.3

As the sides have the same length, we can glue sides by orientation-preserving isometries. To give F a hyperbolic atlas, we use the unique isometries gluing the edges to make charts for the edges. As the angle sum is 2π we get a chart for the vertices.

Remark: The images of the fundamental 4g-gon under the elements of the group generated by the gluing isometries form a tiling of $\mathbb{H}^2$. In the construction the 4g-gon need not be regular; a general criterion for when a 4g-gon generates a tiling was given by Poincare.

Geodesics in F are defined locally, i.e. we can pull back the geodesics in $\mathbb{H}^2$ via the chart maps.

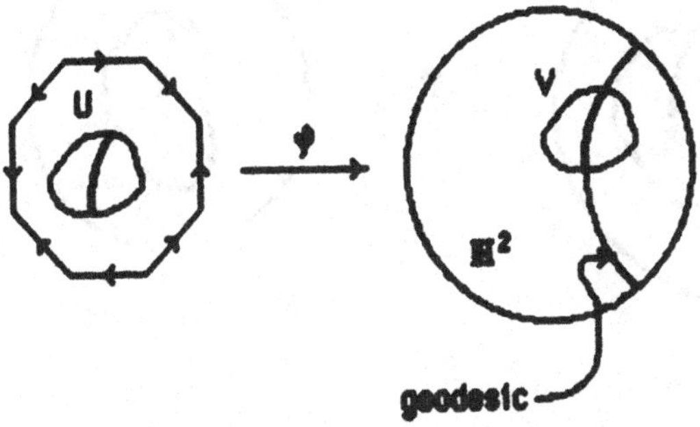

Figure 2.4

Definition: A hyperbolic surface F is *complete* if it is complete as a metric space. (The metric is induced by pulling back the metric on $\mathbb{H}^2$.)

Lemma 2.1: (The easy half of the Hopf–Rinow theorem.) In a complete hyperbolic surface, all geodesics can be extended indefinitely.

Proof: Suppose we have a bounded arc in the surface which is geodesic. Take a sequence of points tending toward an end. This sequence is Cauchy, hence there exists an endpoint. Take a chart with this endpoint as center of the Poincare disk. Now we can extend the geodesic and pull back this extension to F via the chart map. Taking a maximal extension finishes the proof.

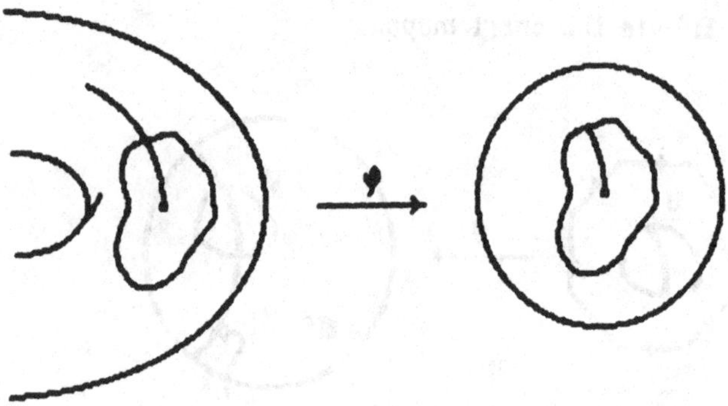

Figure 2.5

Theorem 2.2: Any complete, connected, simply-connected hyperbolic surface is isometric to $\mathbb{H}^2$.

Remark: This implies that the universal cover of a compact hyperbolic surface is $\mathbb{H}^2$.

Proof: We construct maps D, E as indicated below.

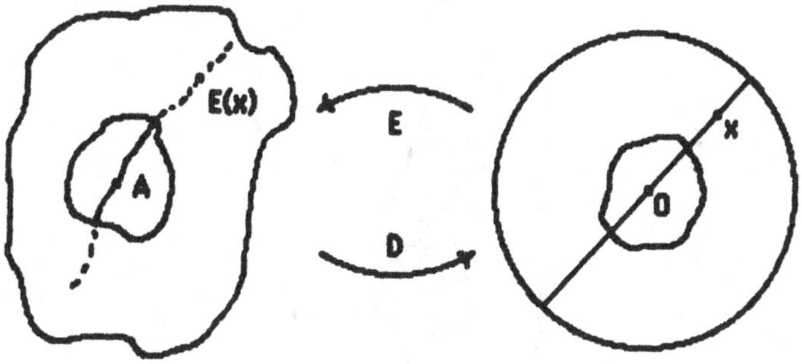

Figure 2.6

E = *Exponential*

Choose A in F and a chart neighborhood $\varphi:U \to \mathbb{H}^2$ such that $\varphi(A)=0$. For $x \in \mathbb{H}^2$ extend the geodesic $\varphi^{-1}(0x)$ to a complete geodesic and define E(x) as the point on the extended geodesic with dist(A, E(x)) = dist(0,x).

D =*Developing*

Claim: There exists a unique map $D:F \to \mathbb{H}^2$ such that

1) D is a local isometry,

2) $D|U = \varphi$.

The map D is constructed by "analytic continuation". Choose a path C from A to B in F. Cover C by a sequence of "round" charts $U=U_0, U_1, \ldots, U_n$ with $\varphi_i:U_i \to \mathbb{H}^2$. Choose points x_i in C such that $[x_i, x_{i+1}] \subseteq U_i$.

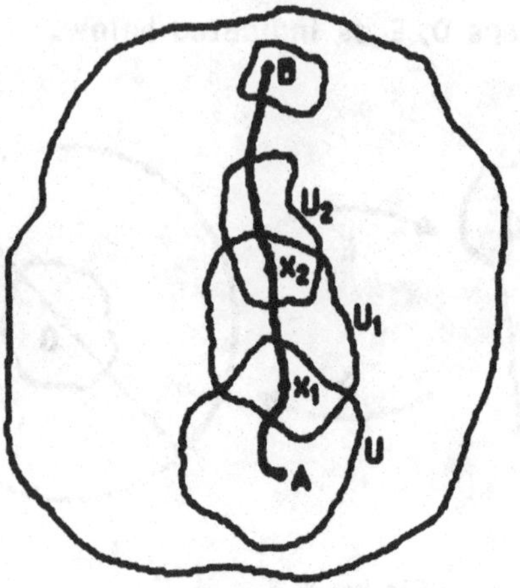

Figure 2.7

Suppose $\varphi_j = \varphi_{j-1}$ on the component of $U_j \cap U_{j-1}$ containing x_j for $j=1, \dots, i$. We wish to perform an induction step, so suppose that $\varphi_{i+1} \neq \varphi_i$ on the component of $U_{i+1} \cap U_i$ containing x_i. Then $\varphi_{i+1} = g_i \varphi_i$ where g_i is an isometry of $\mathbb{H}^2$ uniquely determined by the data. Replace φ_{i+1} by $g_i^{-1}\varphi_{i+1}\colon U_{i+1} \to \mathbb{H}^2$ and now the chart maps agree on the overlaps.

Set $D(B) = \varphi_n(B)$ and note:

(1) By refining coordinate covers, it follows that D(B) depends only on C, and not on U_i, φ_i, or x_i.

(2) If C_1, C_2 are homotopic paths, they define the same value of D(B) as small homotopies do not leave the coordinate cover.

(3) F simply-connected implies that D is well-defined.

Observe that $D \cdot E = 1_{\mathbb{H}^2}$ so $E \cdot D$ is a retraction of F onto the closed subset Im(E) of F. But Im(E) is open by invariance of domain and F is connected, therefore $E \cdot D = 1_F$.

Remark: By applying Theorem 2.2 to the universal cover $\tilde{F}$ of a closed hyperbolic surface F we see that $\tilde{F}$ is isometric to $\mathbb{H}^2$. In fact we shall usually identify $\tilde{F}$ with $\mathbb{H}^2$.

Further, $F = \tilde{F}/\{\text{deck translations}\} \cong \mathbb{H}^2/\Gamma$ where Γ is a subgroup of $PSL_2(\mathbb{R})$ isomorphic to $\pi_1(F)$. As we can lift ϵ-neighborhoods of points to the universal cover, Γ is a discrete subgroup of $PSL_2(\mathbb{R})$. Indeed, this fact also shows that there are no parabolic elements in $\Gamma-\{1\}$. For F compact implies there is a uniform ϵ for all of F, and hence there exists an $\epsilon > 0$ such that for every $g \in \Gamma-\{1\}$ the hyperbolic distance between x and g(x) is greater than ϵ. On the other hand, in the half-plane model a parabolic element is represented by a horizontal translation through a Euclidean distance b. But the metric in this model is given by ds/y, and so points "high up" on the "Y-axis" are moved an arbitrarily small hyperbolic distance.

Notice that as the action of $\pi_1(F)$ on $\mathbb{H}^2$ is free, $\Gamma-\{1\}$ has no elliptic elements. We conclude that every element of $\Gamma-\{1\}$ is hyperbolic.

Remark: This rules out $\mathbb{Z}\oplus\mathbb{Z}$ as a subgroup of $\pi_1(F)$, hence the torus has no hyperbolic structure.

Definition: A closed curve in a surface (or any other space) is *essential* if it is not null-homotopic.

Lemma 2.3: Every essential closed curve in a closed hyperbolic surface is freely homotopic to a unique closed geodesic.

Proof: Let C be an essential closed curve in F. Choose an x on C and an $\tilde{x} \in \tilde{F} = \mathbb{H}^2$ projecting to x. Let $\tilde{C}$ be the complete curve in $\mathbb{H}^2$ containing $\tilde{x}$ and projecting to C. We can find a $g \in \Gamma$ such that the segment $[\tilde{x}, g\tilde{x}]$ goes once around C.

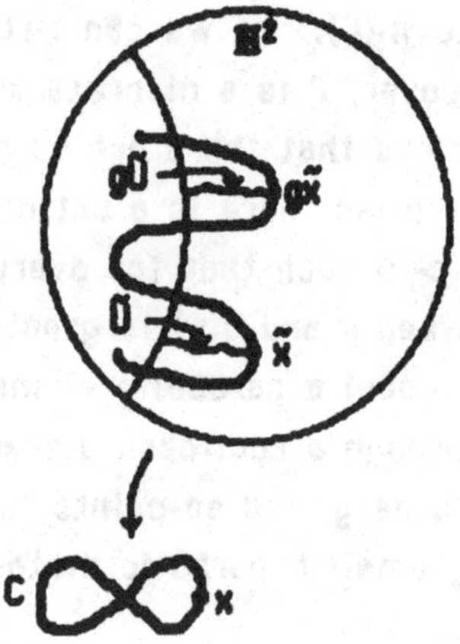

Figure 2.8

As g is hyperbolic, g has a geodesic axis $\tilde{\gamma}$. The image of $\tilde{\gamma}$ in F is a closed geodesic γ. Choose any point $\tilde{y}$ on $\tilde{\gamma}$ and any path $\tilde{U}$ connecting $\tilde{y}$ to $\tilde{x}$. The circuit from $\tilde{y}$ to $\tilde{x}$, along $\tilde{C}$ to $g\tilde{x}$, along $g\tilde{U}$ to

$g(\tilde{y})$, and back to $\tilde{y}$ on $\tilde{\gamma}$ bounds a singular "rectangle" in $\mathbb{H}^2$, which projects to a singular annulus in F. This annulus is a free homotopy from C to γ.

For uniqueness: Let C be an essential closed curve freely homotopic to a closed geodesic γ, with free homotopy f: $S^1 \times I \to F$. As before, let $\tilde{C}$ be a "component" of the preimage of C. Let $\tilde{f}: \mathbb{R} \times I \to \mathbb{H}^2$ cover f such that $\tilde{f}(\mathbb{R} \times 0) = \tilde{C}$. Let $\tilde{\gamma} = \tilde{f}(\mathbb{R} \times 1)$ have endpoints P,Q.

Claim: There exists a $d > 0$ such that $\tilde{f}(\mathbb{R} \times I) \subseteq$ hyperbolic d–neighborhood of $\tilde{\gamma}$.

Proof of claim: Since S^1 is compact, there exists an upper bound d to the hyperbolic lengths of the arcs $f(z \times I)$ for $z \in S^1$. Then in the Euclidean metric on the Poincare disk, $\mathrm{dist}(\tilde{C}, \tilde{\gamma}) \to 0$ as one approaches P or Q. In particular, if C and γ are freely homotopic geodesics, they have the same endpoints and therefore coincide.

Exercise: Recall that a curve in a surface is *simple* if it does not intersect itself. Do there exist simple, non–closed geodesics in a closed hyperbolic surface ?

Definition: A *closed 1–submanifold* of a surface F is a disjoint union of simple closed curves in F. An *essential 1–submanifold* is a closed 1–submanifold in which every component is essential and no two components are homotopic.

Lemma 2.4: Every essential 1–submanifold C of a closed hyperbolic surface is isotopic to a unique geodesic 1–submanifold.

Proof: Each component of C is homotopic to a unique closed geodesic by Lemma 2.3. As no two components of C are homotopic, these geodesics are all distinct; let γ be their union. Note that the preimage of γ in $\mathbb{H}^2$ is obtained from the preimage of C in $\mathbb{H}^2$ by replacing each component with the geodesic having the same endpoints on S^1_∞. Since any two components $\tilde{C}_1$, $\tilde{C}_2$ of the preimage of C are disjoint, their endpoints are not separated on S^1_∞, i.e., if P_1, Q_1 are the endpoints of $\tilde{C}_1$ and P_2, Q_2 are the endpoints of $\tilde{C}_2$, then $P_1 \cup Q_1$ does not separate $P_2 \cup Q_2$. Hence the geodesic P_1Q_1 does not meet the geodesic P_2Q_2. It follows that the preimage of γ is a collection of disjoint geodesics, hence γ is simple. To complete the proof of Lemma 2.4 we need a definition and a lemma.

Definition: Essential 1-submanifolds C_1, C_2 have *minimal intersection* if they intersect transversely and if there do not exist arcs A_1, A_2 in C_1, C_2 respectively, having common endpoints, and such that $A_1 \cup A_2$ is the boundary of a disk in F.

Figure 2.9

Lemma 2.5: If C_1, C_2 are essential 1-submanifolds of a closed surface, then C_2 is isotopic to an essential 1-submanifold having minimal intersection with C_1.

Proof: We may assume that C_1, C_2 are transverse. If they have non-minimal intersection, an innermost disk argument shows that we can find arcs A_1, A_2 such that $\partial A_1 = \partial A_2$ and such that $A_1 \cup A_2$ bounds a disk D with IntD disjoint from $C_1 \cup C_2$. By "pushing C_2 across D" we reduce the number of intersections of C_1, C_2. Continue until C_1, C_2 have minimal intersection.

Returning to the proof of Lemma 2.4, it follows that C may be moved via an isotopy until C and γ have minimal intersection. Choose a component $\tilde{C}$ of the preimage of C in $\mathbb{H}^2$ and examine how it intersects the preimage of γ. Since $C \cap \gamma$ is minimal, $\tilde{C}$ meets no component of the preimage of γ more than once.

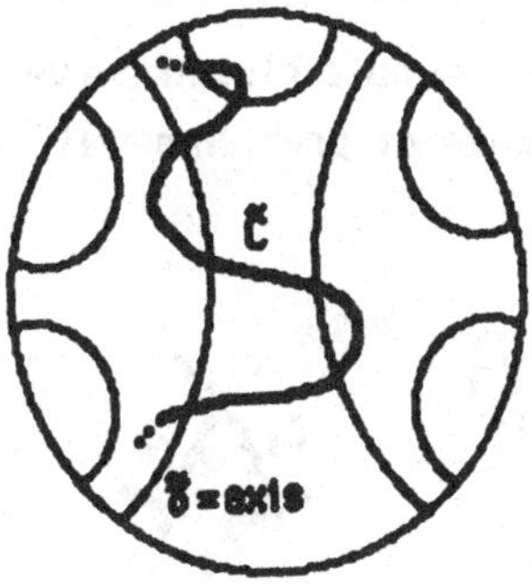

Figure 2.10

Note that $\tilde{C}$ has the same endpoints as a component $\tilde{\gamma}$ of the preimage of γ, and there is a deck translation leaving $\tilde{C}$ and $\tilde{\gamma}$ invariant. If $\tilde{C} \cap \tilde{\gamma}$ is not empty, it contains an an orbit of a hyperbolic isometry and is therefore infinite; but $|\tilde{C} \cap \tilde{\gamma}| \leq 1$, hence $\tilde{C} \cap \tilde{\gamma}$ is empty. If $\tilde{\gamma}'$ is any other component of the preimage of γ, then $|\tilde{C} \cap \tilde{\gamma}'|$ is even, and hence equal to zero. It follows that $C \cap \gamma$ is empty.

Since C and γ are homotopic, they are homologous, i.e. they cobound part of the surface; call it N. Moreover, C homotopic to γ implies that N has genus 0 and is thus an annulus. Therefore, as F is orientable, C and γ are isotopic.

Lemma 2.6: Suppose C_1 and C_2 are transverse essential 1-submanifolds of a hyperbolic surface F and no component of C_1 is isotopic to a component of C_2. Then C_1 and C_2 have minimal intersection if and only if there exists a homeomorphism $h:F \to F$, isotopic to the identity, such that $h(C_1)$ and $h(C_2)$ are both geodesic 1-submanifolds.

Remark: This fails for non-simple closed curves and for three or more 1-submanifolds, because of problems with triple points, see Figure 2.11.

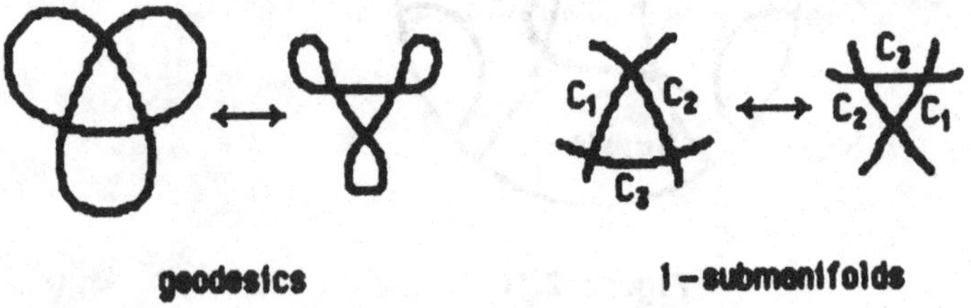

Figure 2.11

Proof: If both C_1, C_2 are geodesic, they have minimal intersection because no component of the preimage of C_1 in the universal cover $\mathbb{H}^2$ meets any component of the preimage of C_2 more than once.

Conversely, suppose C_1, C_2 have minimal intersection. By Lemma 2.4, we can assume that C_2 is geodesic. Again by lemma 2.4, C_1 is isotopic to a geodesic 1-submanifold γ_1. This isotopy can be achieved by a series of "pushes across disks" (Lemma 2.5) followed by a "push across an annulus". We shall choose this isotopy so that it leaves C_2 invariant.

Suppose A and α are arcs of C_1, C_2 respectively such that $A \cup \alpha$ bounds an innermost disk D (see Figure 2.12). Since the 1-submanifold C_2 has minimal intersection with both C_1 and γ_1, it follows that $C_2 \cap D$ is a family of arcs crossing D from "top to bottom". So we can choose our "push across D" to leave C_2 invariant (although not pointwise fixed). After this sequence of pushes, $C_1 \cap \gamma_1$ is empty.

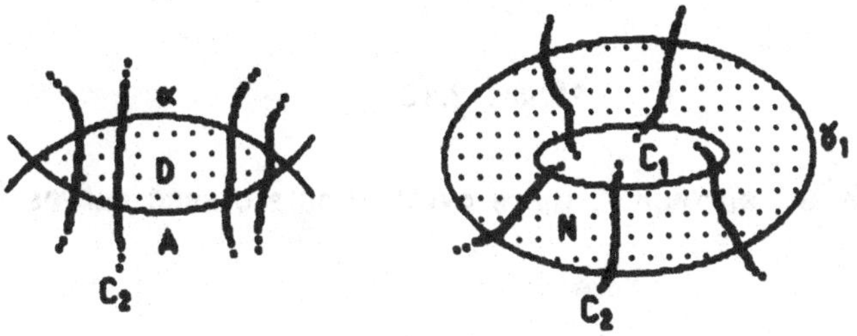

Figure 2.12

Now we can move C_1 to γ_1 via an isotopy which is the identity outside of an annulus N. As above, each component of $C_2 \cap N$ is an arc joining the two boundary components of N and again this isotopy may be chosen to leave C_2 invariant.

Exercise: Let $i(C_1,C_2)$ = the *geometric intesection number* of C_1 and C_2 be the minimum value of $|C_1' \cap C_2'|$ where C_1' is homotopic to C_1. Prove that C_1, C_2 have minimal intersection if and only if $|C_1 \cap C_2| = i(C_1,C_2)$.

Definition: Essential 1-submanifolds C_1, C_2 of F *fill* F if they have minimal intersection and every component of $F-(C_1 \cup C_2)$ is a disk.

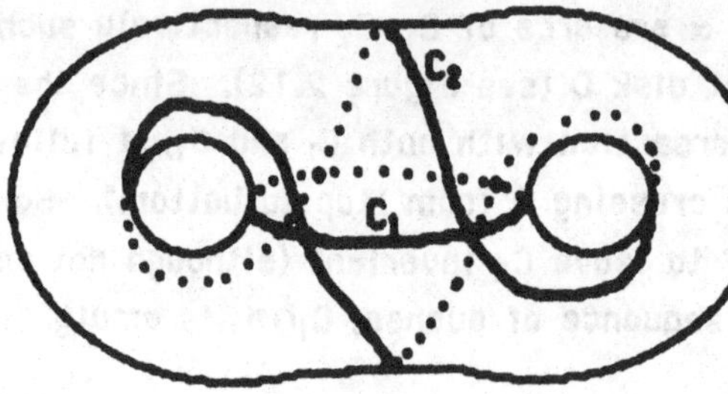

Figure 2.13

Exercise: On any surface F, there exist simple closed curves C_1, C_2 filling F.

Theorem 2.7: Let $h: F \rightarrow F$ be an orientation preserving automorphism of a closed hyperbolic surface F. If for every essential simple closed curve C in F there is an integer $k>0$ such that $h^k(C)$ is homotopic to C, then there exists an integer $n>0$ such that h^n is isotopic to the identity.

Proof: Choose geodesics C_1, C_2 filling F. There exist integers k_1, k_2 such that $h^{k_1}(C_1) \simeq C_1$ and $h^{k_2}(C_2) \simeq C_2$. Setting $k = k_1k_2$, $h^k(C_1) \simeq C_1$

for $i=1, 2$. The curves $h^k(C_1)$ and $h^k(C_2)$ have minimal intersection, so by Lemma 2.6 there is a homeomorphism g, isotopic to the identity, such that $g \cdot h^k(C_i)$ is a closed geodesic for $i=1,2$. It follows that $g \cdot h^k(C_i) = C_i$ for $i = 1,2$, so $g \cdot h^k(C_1 \cap C_2) = C_1 \cap C_2$.

Considering $C_1 \cup C_2$ as a graph on F, $g \cdot h^k$ permutes the vertices of this graph, so there exists an $m \geq 0$ such that $(g \cdot h^k)^m$ is isotopic to a homeomorphism $f\colon F \to F$ which restricts to the identity on $C_1 \cup C_2$. As g is isotopic to the identity, it follows that h^{km} is isotopic to f. As the complementary regions of $C_1 \cup C_2$ are disks, the Alexander trick shows that f is isotopic to the identity (see Figure 2.14).

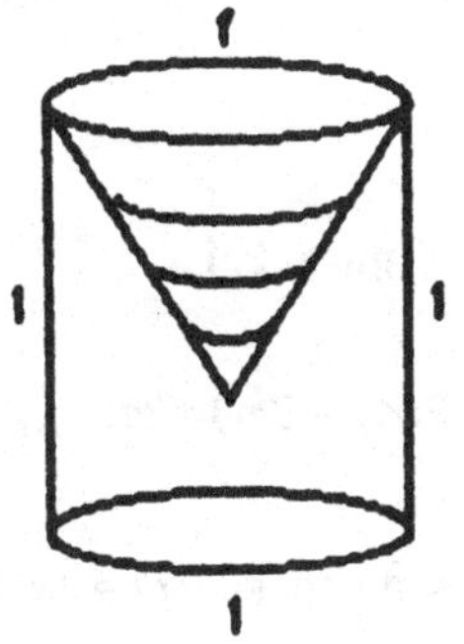

Figure 2.14

The remainder of this chapter is devoted to the study of complete hyperbolic surfaces with finite area and geodesic boundary. One result is that there is a lower bound for the area of such surfaces. We begin by studying compact surfaces.

Example: Choose a right rectangular geodesic hexagon in $\mathbb{H}^2$. Take a second copy and "glue" every other edge (see Figure 2.15). The resulting surface is the hyperbolic "pair of pants".

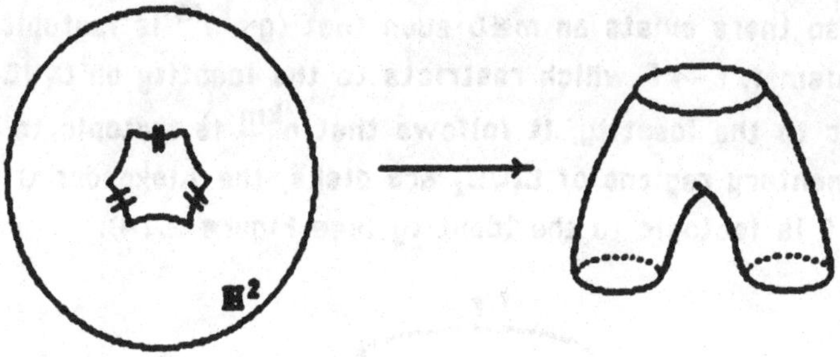

Figure 2.15

The area of this surface is $2(4\pi-3\pi)=2\pi$, $\chi_F=-1$.

Lemma 2.8: A compact hyperbolic surface with geodesic boundary has area $-2\pi\chi_F$.

Proof: First we show that every closed hyperbolic surface is made by gluing the edges of a geodesic polygon in pairs. As usual $F = \mathbb{H}^2/\Gamma$, where Γ is a discrete group of isometries.

Choose $P \in \mathbb{H}^2$ and for each $g \in \Gamma$ let $U_g = \{x \in \mathbb{H}^2 |\ d(x,P) \leq d(x,gP)\}$, a closed hyperbolic half-plane with geodesic boundary. Let U be the intersection over all $g \in \Gamma$ of the sets U_g. Note that U is closed, and for all $x \in \mathbb{H}^2$ there exists an element $g \in \Gamma$ such that $g(x) \in U$. As F is compact it has a diameter d such that every point of $\mathbb{H}^2$ is within distance d of gP for some $g \in \Gamma$. It follows that every point of U is within distance d of P.

Clearly FrU, the frontier of U, is contained in the union of the sets $Fr(U_g)$. Since $\{gP \mid g \in \Gamma\}$ is discrete there are only finitely elements $g \in \Gamma$ such that $d(P,gP) \leq 2d$. Hence U meets only finitely many of the sets $Fr(U_g)$, implying that FrU is a finite collection of geodesic arcs. Call U a *Poincare polygon* for F. Since each point gP has a Poincare polygon the edges of U are identified in pairs.

Suppose U has 2e edges, then this gives rise to a cell decomposition of F with a single 2-cell, e edges, and some number v of vertices. The Euler characteristic of F is $\chi_F = v-e+1$. So $e = v-\chi_F+1$ and $2e = 2(v-\chi_F+1)$.

By the Gauss-Bonnet theorem the area of our 2e-gon U is $2(v-\chi_F)\pi - 2\pi v = -2\pi\chi_F$. Therefore the area of the closed hyperbolic surface F is $-2\pi\chi_F$.

If F has non-empty geodesic boundary ∂F, the double DF of F has a natural hyperbolic structure. Clearly Area(DF) = 2Area(F) and, as F is compact, $\chi_{DF} = 2\chi_F$. Since DF is closed the lemma follows.

Lemma 2.9: An unbounded complete hyperbolic surface with finite area is homeomorphic to a closed surface less a finite set and has area $-2\chi_F$.

Example: Consider an ideal triangle, Area = π. Double it to get a three punctured sphere, Area=2π.

Figure 2.16

Proof of 2.9: Since F is complete, F ≅ $\mathbb{H}^2/\Gamma$ with Γ discrete. Note that there may be non-hyperbolic elements in Γ.

As in the previous proof let $U_g = \{x \in \mathbb{H}^2 | \ d(x,P) \leq d(x,gP)\}$, and let U be the intersection of all the half-planes U_g. Again FrU is contained in the union of the sets $Fr(U_g)$. Although U is no longer compact, as illustrated in Figure 2.17, the argument of Lemma 2.8 shows that FrU is a locally finite union of geodesic arcs.

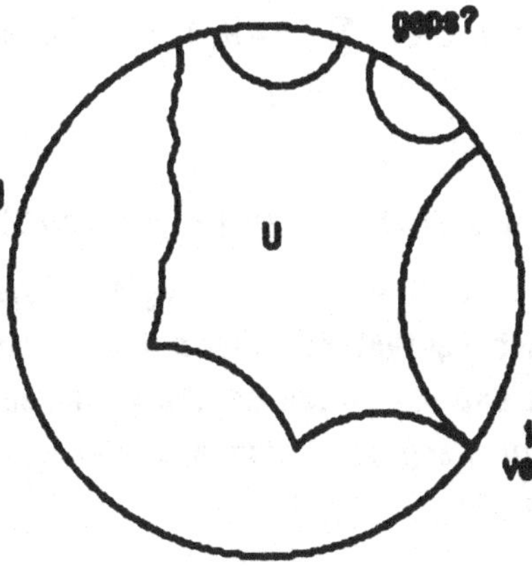

Figure 2.17

If the Euclidean closure of U in the Poincare disk has n distinct points on S^1_∞, then U contains an ideal polygon with n vertices, thus $(n-2)\pi <$ area(U). It follows that the Euclidean closure of U has only finitely many points on S^1_∞.

Let α_v be the angle at the vertex v of U. Vertices occur when v is equidistant from two or more distinct translates of P, so v is the intersection of distinct geodesics, and hence $\alpha_v < \pi$. Note that as one can take a finite number of interior vertices of U and make a finite polygon G contained in U. Area(G) = $(n-2)\pi-$(angle sum of G) and so $n\pi-$(angle sum of G) = 2π + area(G), implying that

$$\sum_{v \in G}(\pi-\alpha_v) \leq 2\pi + \text{area}(G).$$

From this one concludes that $\sum_{v \in \mathrm{Int}U} (\pi-\alpha_v) \leq 2\pi + \mathrm{area}(U)$.

Let $A = \{v \mid \alpha_v \leq 2\pi/3\}$, $B = \{v \mid \alpha_v > 2\pi/3\}$. Since $\sum(\pi-\alpha_v)$ converges, A is finite. Call vertices v_1, v_2 equivalent if they have the same image in F. Each equivalence class has no more than two B-vertices and at least one A-vertex as the angle sum is 2π. Hence there are only finitely many equivalence classes, and thus U has only finitely many vertices.

Performing the identifications shows that F is made from U∪{ideal vertices} by identifying edges in pairs and then deleting the ideal vertices. It follows that F is homeomorphic to a closed surface less a finite set of points.

Now F is a cell complex with some vertices deleted; one 2-cell, e edges, and v surviving vertices.

$$\begin{aligned} \mathrm{Area}(U) &= (n-2)\pi - \text{angle sum} \\ &= (2e-2)\pi - 2\pi v \text{ as ideal vertices have angle } 0 \\ &= (2e-2-2v)\pi \\ &= -2\pi\chi_F. \end{aligned}$$

Remark: If $F = \mathbb{H}^2/\Gamma$ is of finite area but non-compact, then Γ has parabolic elements.

Example: A 1-punctured torus, obtained from an ideal quadrilateral, area = 2π.

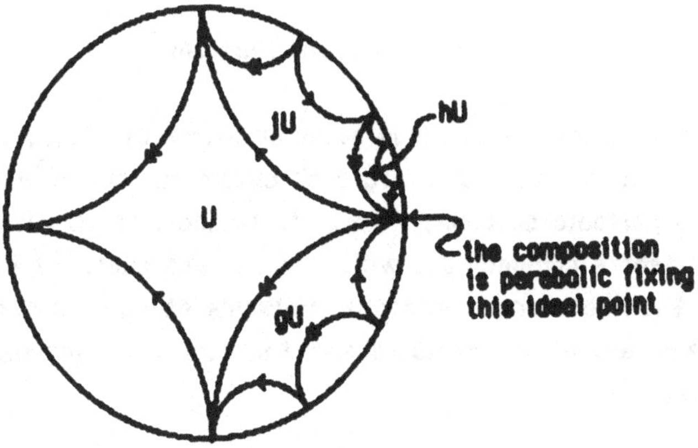

Figure 2.18

Theorem 2.10: A complete hyperbolic surface F with finite area and geodesic boundary is homeomorphic to a compact surface less a finite set and has area $-2\pi\chi_F + \pi\chi_{\partial F}$.

Proof: Double F; Area(DF) = 2 Area(F) and $\chi_{DF} = \chi_F + \chi_F - \chi_{\partial F}$.

Corollary 2.10.1: The area of any complete hyperbolic surface F with totally geodesic boundary is $n\pi$, $n \geq 1$. In particular π is a lower bound for such areas.

§3. Geodesic Laminations

Our principal goal is to study automorphisms of closed surfaces up to isotopy. It turns out that every such automorphism is either isotopic to a periodic automorphism or is isotopic to an automorphism leaving invariant a compact set which can be expressed as a union of geodesics. Such sets are implicit in Nielsen's study of surface automorphisms and play a basic role in Thurston's refinement of Nielsen's ideas.

Definition: For a closed hyperbolic surface F, a *geodesic* in F is the image of a complete geodesic in $\mathbb{H}^2 \cong \tilde{F}$. A geodesic in F is *simple* if it has no transverse self intersections.

A *(geodesic) lamination* on F is a non-empty closed subset L of F which is a disjoint union of simple geodesics. We will prove that such an L is a union of geodesics in just one way. The geodesics contained in L are called the *leaves* of L.

Examples: Finitely many disjoint closed simple geodesics form a lamination.

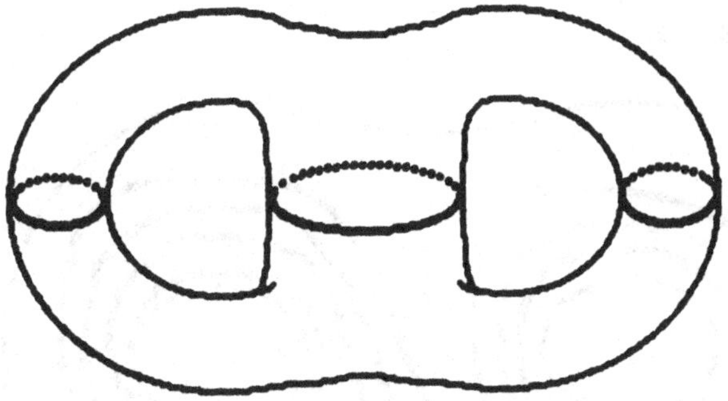

Figure 3.1

Slightly less trivial examples are obtained by including infinite leaves spiraling towards closed leaves.

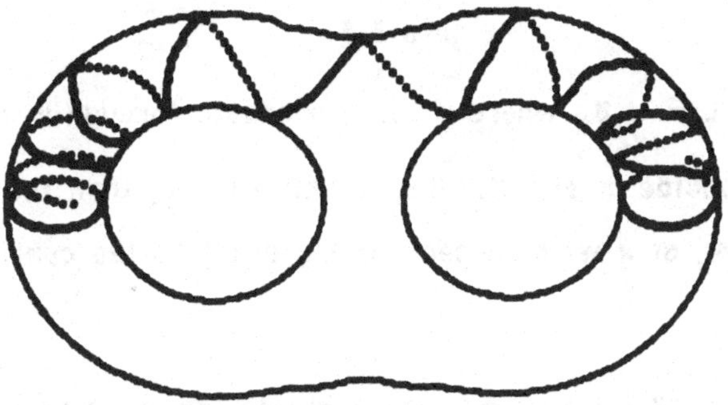

Figure 3.2

The most interesting examples will not be of either of these types but rather will be "limits" of long simple closed geodesics. Usually these laminations will not contain any closed leaves. A simple closed geodesic approximating such a lamination might look like the one shown in Figure 3.3.

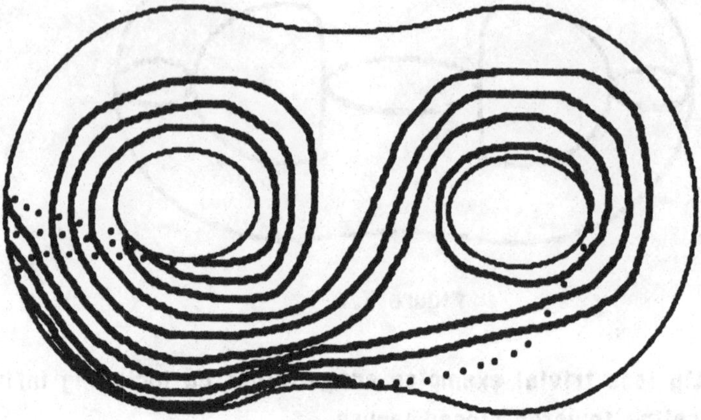

Figure 3.3

Lemma 3.1: If $L = \bigcup_{x \in L} \gamma_x$ where γ_x is a geodesic through x, and γ_x, γ_y either coincide or are disjoint for all $x, y \in L$, then the direction (see below) of γ_x at x with respect to any chart varies continuously with x.

Proof: Let $\varphi: U \to \mathbb{H}^2 = \text{Int } D^2 \subseteq \mathbb{R}^2$ be a chart, and let ℓ be a fixed line in $\mathbb{R}^2$. Let x be a point in U and let γ be a geodesic through x. Define the *direction of* γ *at* x (with respect to φ and ℓ) as the angle between $\varphi(\gamma)$ and the Euclidean parallel to ℓ through $\varphi(x)$.

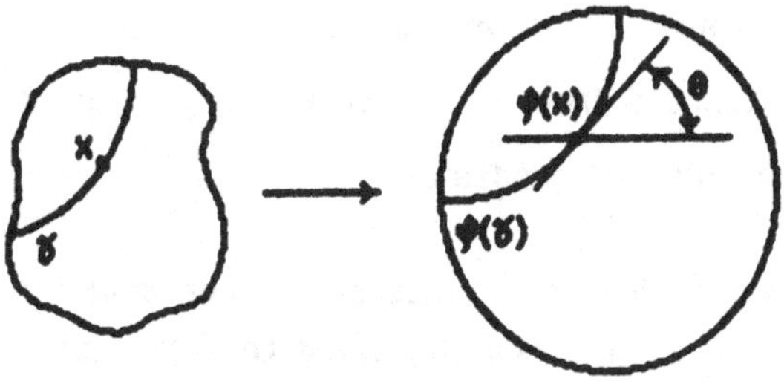

θ = direction of γ at x

Figure 3.4

Now just observe that geodesics will intersect if the directions are far apart. The lemma is now an easy exercise.

Lemma 3.2: The closure of a non-empty disjoint union L of geodesics is a lamination.

Proof: We must show $\bar{L}$ is a disjoint union of simple geodesics. Let x_n be a sequence of points in L converging to a point x in F. Let γ_n be the geodesic (in the given decomposition of L into simple geodesics) containing x_n.

Observe that the set of directions of geodesics at x is compact, indeed, homeomorphic to $\mathbb{R}P^1$. By passing to subsequence we may ensure that the direction of γ_n at x_n converges as $n \to \infty$. (Here directions are measured in some fixed chart about x.) Let γ be the geodesic through x with this limiting direction.

If y is a point on γ at (signed) distance d from x, let y_n be the point on γ_n at distance d from x_n. Then $y_n \to y$, hence γ is contained in $\bar{L}$, and so $\bar{L}$ is a union of geodesics.

If two geodesics $\gamma, \gamma' \subseteq \bar{L}$ intersect transversely at $x \in F$, we can find geodesics β, β' in L passing close to x and approximating arbitrarily closely the directions of γ, γ' at x. When the approximation is sufficiently close, β and β' must intersect near x. A similiar argument shows that the leaves of $\bar{L}$ are simple.

Remark: The geodesics which are members of a lamination are considered unoriented. This is necessary, for if one forms the lamination which "limits" the "sequence" of longer and longer simple closed curve geodesics, it must be unoriented as the example below shows.

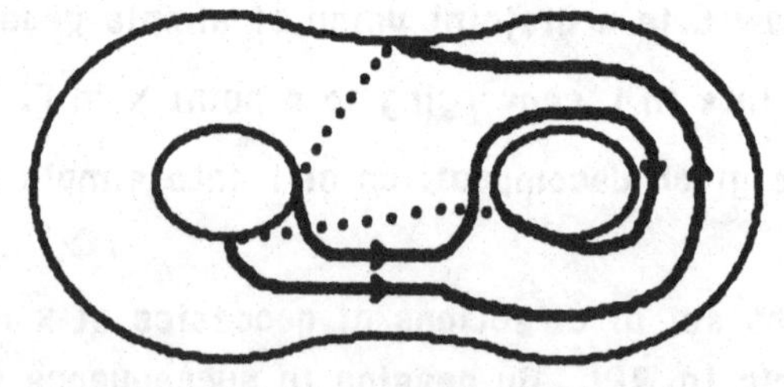

Figure 3.5

Lemma 3.3: A geodesic lamination in a closed orientable hyperbolic surface F is nowhere dense in F, and can be expressed as a union of geodesics in just one way.

Proof: Since F is closed and hyperbolic, it has strictly negative Euler characteristic, so there is no continuously varying line field defined on all of F. It follows that any lamination L in F is a proper subset of F. We shall show that $\tilde{L}$, the preimage of L in $\mathbb{H}^2$, can be expressed as a disjoint union of geodesics in just one way.

Let $\tilde{L} = \bigcup_{x \in \tilde{L}} \gamma_x$ where γ_x is a geodesic through x, and γ_x, γ_y either coincide or are disjoint for all $x, y \in L$. Suppose that $\tilde{L}$ contains a geodesic arc α meeting γ_x transversely for some point x in the interior of α. Define $\Phi : \alpha \times \mathbb{R} \to \mathbb{H}^2$ to be the map sending (y,t) to the point on γ_y at distance t from y. (Here we have chosen an orientation on the normal to α.) Lemma 3.1 implies that Φ is continuous. Notice that $\Phi(\alpha \times \mathbb{R})$ is contained in $\tilde{L}$. Given $d > 0$, there exists a point z on γ_x such that the hyperbolic d-neighborhood U of z is contained in $\Phi(\alpha \times \mathbb{R})$ (see Figure 3.6).

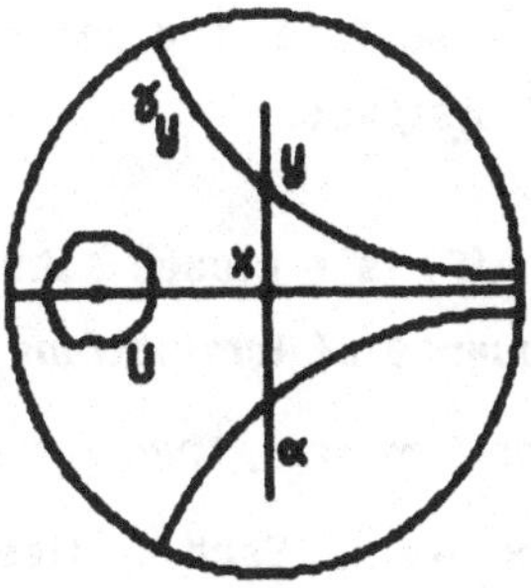

Figure 3.6

But if d is the diameter of a Poincare polygon for F, then U maps onto F, implying that L = F, a contradiction. It follows that L has empty interior; as L is closed, it is nowhere dense.

Definition: Suppose that A,B are closed subsets of a compact metric space X. The *Hausdorff distance* d(A,B) is defined by

$$d(A,B)\leq\epsilon \text{ if and only if } A \subseteq N_\epsilon(B) \text{ and } B \subseteq N_\epsilon(A).$$

Exercises: Show that:

1) This defines a metric on the set 2^X of non-empty closed subsets of X.

2) The topology on 2^X depends only on the topology of X as any two metrics on X are equivalent.

3) The space 2^X is totally bounded and complete, therefore compact.

Theorem 3.4: Let $\Lambda = \Lambda(F)$ be the set of all geodesic laminations on a closed orientable hyperbolic surface. The set $\Lambda(F)$ has a compact metric defined by Hausdorff distance.

Proof: We must show that $\Lambda(F)$ is a closed subset of the compact space 2^X. If (L_n) is a sequence of laminations converging to a compact set L in the Hausdorff metric, then for each $x \in L$ we can find nearby points $x_n \in L_n$ with $x_n \to x$. Each x_n lies on a leaf γ_n of L_n and by passing to a subsequence we can ensure that γ_n tends to a geodesic

γ through x. But, if we choose a different subsequence we may arrive at a limit geodesic with a different direction at x. So to be sure that L is a lamination we must control the directions of geodesics as well.

Consider the *projective tangent bundle* $PT(F) = \{(x,\sigma) \mid x \in F,\ \sigma$ a closed unoriented geodesic segment of length 2, say, centered on $x\}$. Topologize $PT(F)$ with charts:

$$\begin{array}{ccccccccc} (x,\sigma) & \in & PT(F) & \supset & p^{-1}(U) & \cong & & U \times \mathbb{R}P^1 \\ \downarrow p & & \downarrow p & & \downarrow p & & & \downarrow \text{projection} \\ x & \in & F & \supset & U & \xrightarrow{1_U} & & U \end{array}$$

i.e., take the collection of p^{-1}(chart neighborhoods) as a basis for the topology. Note that $PT(F)$ is a compact 3-manifold and the map $p: PT(F) \to F$ is continuous. Moreover, $2^{PT(F)}$ is compact and metrizable as $PT(F)$ is compact and metrizable.

Given γ a geodesic in F, we have the (lifted) geodesic $\hat{\gamma} = \{(x,\sigma) \mid \sigma \subseteq \gamma\}$ in $PT(F)$. Notice that every point in $PT(F)$ lies on a unique lifted geodesic. For L a lamination in F, define $\hat{L} = \{\hat{\gamma} \mid \gamma \in L\}$; $\hat{\Lambda} = \{\hat{L} \mid L \in \Lambda(F)\}$.

Now we have:

$$\begin{array}{ccc} \hat{L} & \subset & PT(F) \\ {\scriptstyle 1-1 \atop \scriptstyle \text{onto}}\downarrow p & & \downarrow p \\ L & \subset & F \end{array}.$$

Lemma 3.1 implies that p^{-1} is continuous on L, so p is a homeomorphism, hence $\hat{L}$ is compact. This shows that the lifted laminations are in $2^{PT(F)}$ and hence $\hat{\Lambda} \subseteq 2^{PT(F)}$. In the diagram below, $p_*: \hat{\Lambda} \to \Lambda$ is 1-1, onto, and continuous.

$$\begin{array}{ccc} \hat{\Lambda} & \subset & 2^{PT(F)} \\ p_* \downarrow & & \downarrow p_* \\ \Lambda & \subset & 2^F \end{array}$$

If we can show that $\hat{\Lambda}$ is closed in $2^{PT(F)}$, it will follow that $\hat{\Lambda}$ is compact, implying that Λ is compact. So suppose that $\hat{L}_n \in \hat{\Lambda}$ and that $\hat{L}_n \to A \in 2^{PT(F)}$. We wish to show that $A = \hat{L}$ for some lamination L in F. Note that $L = p(A)$ is non-empty and compact.

If $x \in L$, $x=p(a)$ where $a = (x,\sigma) \in A \subseteq PT(F)$ (recall that σ is a length 2 geodesic segment centered on x). There exists $(x_n,\sigma_n) \in \hat{L}_n$ converging to (x,σ). Then $x_n \to x$ and the direction of σ_n at x_n converges to the direction of σ at x. As in Lemma 3.2, the prolongation γ of σ is contained in L, and L is a disjoint union of geodesics.

Lemma 3.5: Hausdorff distance on 2^F and $2^{PT(F)}$ define the same topology on $\Lambda(F)$.

Proof: The above argument shows that $p_*: \hat{\Lambda}(F) \to \Lambda(F)$ is a homeomorphism.

Definition: A leaf γ of a lamination L is *isolated* if for each $x \in \gamma$ there exists a neighborhood U of x such that $(U, U \cap L)$ is homeomorphic to (disk, diameter).

Exercise: This holds for every $x \in \gamma$ if and only if it holds for some $x \in \gamma$.

Set $L' = L -$ {isolated leaves}. Note that L' is closed; if L' is not empty we call L' the *derived lamination* of L.

Lemma 3.6: If L' is empty, then L is a finite union of simple closed geodesics and L is an isolated point in $\Lambda(F)$.

Proof: If L' is empty, all the leaves of L are isolated, so L is a closed 1-submanifold of F. The lamination L is a disjoint union of simple closed geodesics. There exists $\epsilon > 0$ such that $\overline{N_\epsilon(L)}$ is a disjoint union of annuli. Suppose L_* is a lamination such that $d(L_*, L) < \epsilon$. Any leaf γ of L_* is contained in $N_\epsilon(L)$ and therefore in $N_\epsilon(C)$ for some closed geodesic C in L. Thus some lift of γ is in $N_\epsilon(\tilde{C}) \subset \mathbb{H}^2$. But the only geodesic in this neighborhood is $\tilde{C}$ itself, so $\gamma = C \subset L$. It follows that $L_* \subseteq L$; since every component of $N_\epsilon(L)$ contains a point of L_*, $L_* = L$.

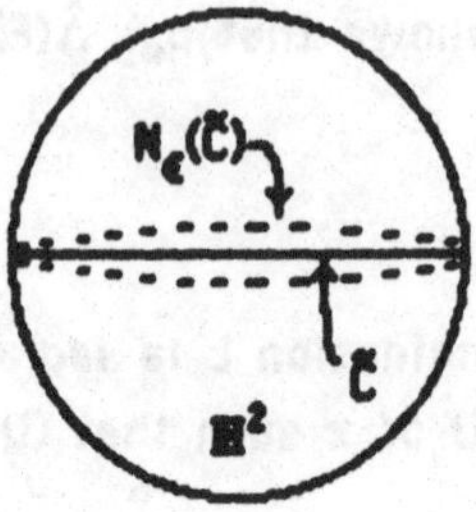

Figure 3.7

Exercises:

1) Is every isolated point of $\Lambda(F)$ a 1-dimensional submanifold? In particular, is the lamination of Figure 3.2 an isolated point of $\Lambda(F)$?

2) Can you find a *perfect* lamination? i.e. a lamination L such that $L' = L$?

Lemma 3.7: Let $h: F_1 \to F_2$ be a homeomorphism of closed orientable hyperbolic surfaces, and let $\tilde{h}: \mathbb{H}^2 \to \mathbb{H}^2$ be a lift of h to the universal cover. Then $\tilde{h}$ has a unique continuous extension over $\mathbb{H}^2 \cup S^1_\infty$.

Remarks:

1) $\mathbb{H}^2 \cup S^1_\infty$ is topologized as the closure of the Poincare disk in the Euclidean plane.

2) This theorem is true in higher dimensions as well.

Proof: We already know that isometries extend over S^1_∞ as they are uniformly continuous with respect to the Euclidean metric on the Poincare disk.

We shall show that for any geodesic γ in $\mathbb{H}^2$, $\tilde{h}(\gamma)$ converges to a point on S^1_∞. Without loss of generality, we let $\tilde{h}(0) = 0$.

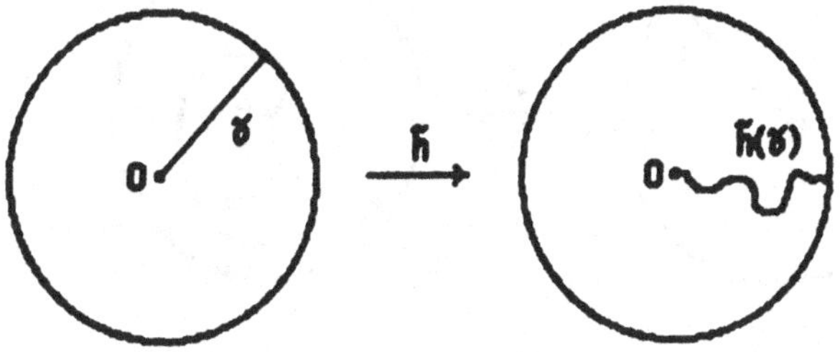

Figure 3.8

Notice that as $\tilde{h}$, $\tilde{h}^{-1}$ are lifts of continuous maps of closed hyperbolic surfaces, they are uniformly continuous with repect to the hyperbolic metric. Hence there exists an integer $k > 0$ such that

$$d(x,y) \leq 1/k \Rightarrow d(\tilde{h}x,\tilde{h}y) \leq 1 \qquad \text{and}$$
$$d(\tilde{h}x,\tilde{h}y) \leq 1/k \Rightarrow d(x,y) \leq 1$$

By subdividing the geodesic joining x and y into k equal subintervals, it follows that:

$$d(x,y) \leq 1 \Rightarrow d(\tilde{h}x,\tilde{h}y) \leq k \qquad (1)$$

Similiarly, for any integer $n > 0$:

$$d(\tilde{h}x,\tilde{h}y) \leq n \Rightarrow d(x,y) \leq kn. \qquad (2)$$

The crucial point is to rule out "spiraling":

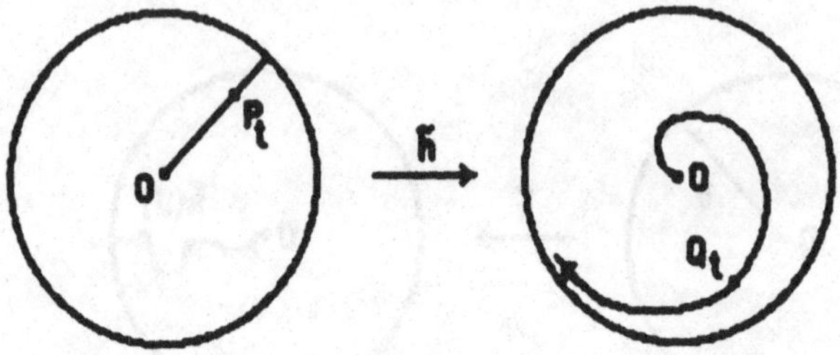

Figure 3.9

Let γ be an oriented geodesic through O, let P_t be the point on γ a distance t from O, and let $Q_t = \tilde{h}(P_t)$. Pick a fixed reference line through O and let θ_t be the angle between OQ_t and this reference line. We want θ_t to have a limit as $t \to \infty$.

Every point on the arc Q_t, Q_{t+1} has distance $\geq t/k$ from O by (2). By (1), $d(Q_t, Q_{t+1}) \leq k$, therefore every point on the geodesic arc Q_tQ_{t+1} has distance $\geq t/k - k$ from O. This distance is at least $t/2k$ if $t \geq 2k^2$.

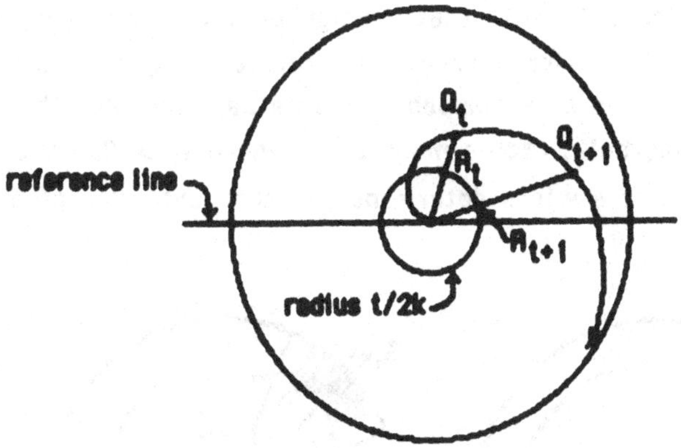

Figure 3.10

Let R_t be the intersection of the ray OQ_t with the circle at center O and hyperbolic radius $t/2k$. Observe that radial projection from the geodesic Q_tQ_{t+1} onto the arc R_tR_{t+1} decreases hyperbolic distance (see Figure 3.10). It follows that:

$$k \geq d(Q_t,Q_{t+1}) \geq \text{arc } R_t,R_{t+1} = |\theta_{t+1}-\theta_t|\sinh(t/2k)$$

Since $\sinh x = (e^x - e^{-x})/2 \geq e^x/4$ if $x \geq 1$. We obtain

$$|\theta_{t+1}-\theta_t| \leq 4ke^{-t/2k} \quad \text{if } t \geq 2k^2.$$

For $u \geq t \geq 2k^2$, $|\theta_u - \theta_t| \leq \int_{t-1}^{u} 4ke^{-s/2k}\,ds = Ce^{-t/2k}$ where C is a function of k only. Hence $\lim_{t\to\infty} \theta_t$ exists, call it θ. We can therefore define our extension of $\tilde{h}$ by taking the endpoint X of γ to the point on S^1_∞ at angle θ from the reference line through O.

Is this continuous? Note that a neighborhood of $\tilde{h}(x)$ can be thought of as a 2ϵ-section exterior of a circle of hyperbolic radius ρ. To find a neighborhood of x mapped into this set, consider the exterior of a circle of hyperbolic radius $\max(k\rho, t_0)$ where $t_0 > 2k^2$ and $Ce^{-t_0/2k} < \epsilon/3$ and a width determined by the continuity of h at A (see Figure 3.11).

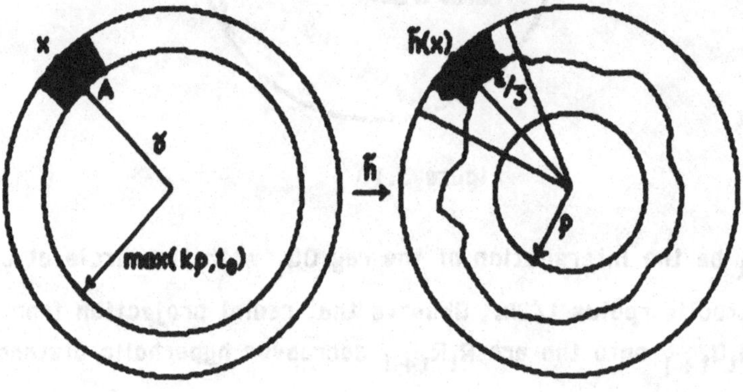

Figure 3.11

Note: $k\rho$ insures that the image of the circle is outside of a circle of radius ρ.

Remarks:

1) We use the same symbol $\tilde{h}$ to denote the unique continuous extension of $\tilde{h}$ over $\mathbb{H}^2 \cup S^1_\infty$ and, by abusing notation, the restriction of this extention to S^1_∞.

2) With a bit more work the hypothesis can be weakened to h a homotopy equivalence.

3) In dimension $n \geq 3$ this is the first step in Mostow's Rigidity Theorem. In that argument one concludes that h is homotopic to an isometry. Although any two ideal triangles in $\mathbb{H}^2$ are isometric, not all ideal simplices in $\mathbb{H}^n$ ($n \geq 3$) have the same volume. However, for all n any two ideal simplices of maximal volume are isometric. Gromov proved Mostow's theorem by showing that the lift of a homotopy equivalence carries maximal volume simplices to maximal volume simplices. By applying this to the collection of simplices obtained from a maximal volume simplex by repeated reflections in faces, he showed that the lift of any homotopy equivalence agrees with an isometry on S_∞^{n-1}.

Lemma 3.8: If $h_0, h_1: F_1 \to F_2$ are homotopic homeomorphisms of closed oriented hyperbolic surfaces and $\tilde{h}_0$ is a lift of h_0, then there is a lift $\tilde{h}_1$ of h_1 such that $\tilde{h}_0 = \tilde{h}_1$ on S^1_∞.

Proof: Let $H: F_1 \times I \to F_2$ be a homotopy between h_0 and h_1. Let $\tilde{H}$ be the lift of H such that $\tilde{H}_0 = \tilde{h}_0$. As H is uniformly continuous with respect to the hyperbolic metric, it follows that the hyperbolic lengths of the arcs $\tilde{H}(a \times I)$ are bounded. Therefore the Euclidean distance between $\tilde{h}_1(x)$ and $\tilde{h}_2(x)$) tends to 0 as x tends to S^1_∞. Hence $\tilde{h}_1(y) = \tilde{h}_2(y)$ for all $y \in S^1_\infty$.

Open Question: Does Aut(F) act "naturally" on F? More precisely, does the projection Φ:Homeo(F) $\rightarrow \pi_0$(Homeo$_+$(F)) = Aut(F) have a 1-sided inverse homomorphism Ψ:Aut(F) $\rightarrow$ Homeo(F) such that $\Phi \cdot \Psi = 1_{\mathrm{Aut}(F)}$? Here are some partial solutions.

1) If $\pi \subseteq$ Aut(F) is finite there is a group homomorphism $\Psi:\pi \rightarrow$ Homeo(F) such that $\Phi \cdot \Psi$. This was proved by Nielsen and Fenchel in the case that π is cyclic or, more generally, solvable. Kerckhoff proved this result for any finite group π by an argument that used Thurston's compactification of Teichmueller space.

2) Morita proved that there is no homomorphism Θ from Aut(F) to Diffeo(F) such that $\Phi \cdot \Theta = 1_{\mathrm{Aut}(F)}$.

3) Gromov and Cheeger proved that Aut(F) acts naturally on the *unit tangent bundle* UT(F). Here UT(F) is a double cover of PT(F) and can be defined as $\{(x,\sigma) \mid \sigma$ an *oriented* geodesic segment of length 2 centered on $x\}$.

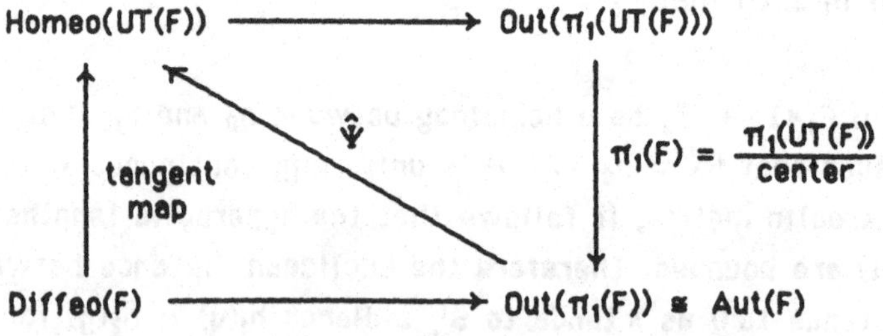

The map $\hat{\Psi}$ is a group homomorphism which gives a canonical self-homeomorphism of UT(F) from each automorphism of F. Gromov's version of Cheeger's construction is given as Theorem 3.10.

Lemma 3.9: If $F = \mathbb{H}^2/\Gamma$ is a closed complete orientable hyperbolic surface, then $UT(F) = Y/\Gamma$ where $Y = \{(a,b,c) \mid a,b,c$ are distinct points of S^1_∞ in counter-clockwise order$\}$.

Proof: Given (a,b,c) in Y, let x be the foot of the perpendicular from c to the geodesic ab, let σ be the length 2 segment of ab centered on x and set $q(a,b,c) = (p(x), \sigma)$.

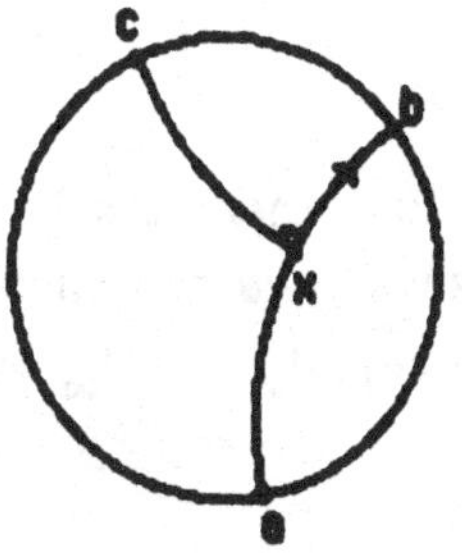

Figure 3.12

This defines a continuous surjection from Y to UT(F), inducing a homeomorphism from Y/Γ onto UT(F).

Theorem 3.10: (Cheeger, Gromov) Any orientation-preserving homeomorphism $h:F_1 \to F_2$ of closed oriented hyperbolic surfaces induces a homeomorphism $\hat{h}:UT(F_1) \to UT(F_2)$. If h is homotopic to k then $\hat{h} = \hat{k}$. If $h_1: F_1 \to F_2$ and $h_2: F_2 \to F_3$ are homeomorphisms then $\widehat{h_2h_1} = \hat{h}_2 \cdot \hat{h}_1$. Finally, $\hat{h}$ carries lifted geodesics to lifted geodesics.

Proof: As usual let $F_i = \mathbb{H}^2/\Gamma_i$ and let $\tilde{h}:\mathbb{H}^2 \to \mathbb{H}^2$ be a lift of $h:F_1 \to F_2$. By Theorem 3.7 we can extend and restrict this map to an orientation-preserving map $\tilde{h}:S^1_\infty \to S^1_\infty$. Let $\bar{h}:Y \to Y$ be the map induced by $\tilde{h}$.

If $g_1 \in \Gamma_1$, then $\tilde{h}g_1 = g_2\tilde{h}$ where $g_2 = h_*(g_1) \in \Gamma_2$. It follows that $\bar{h}$ induces a map $\hat{h}:Y/\Gamma_1 \to Y/\Gamma_2$ which is independent of the choice of the lift of h. Applying Lemma 3.9, we obtain an induced map $\hat{h}: UT(F_1) \to UT(F_2)$.

If h is homotopic to k then there are lifts $\tilde{h}$, $\tilde{k}$ such that $\bar{h} = \bar{k}$, implying that $\hat{h} = \hat{k}$. Moreover, we can choose h_2h_1 to be equal to $\tilde{h}_2 \cdot \tilde{h}_1$, showing that $\overline{h_2h_1} = \bar{h}_2 \cdot \bar{h}_1 : S^1_\infty \to S^1_\infty$ and that $\widehat{h_2h_1} = \hat{h}_2 \cdot \hat{h}_1$.

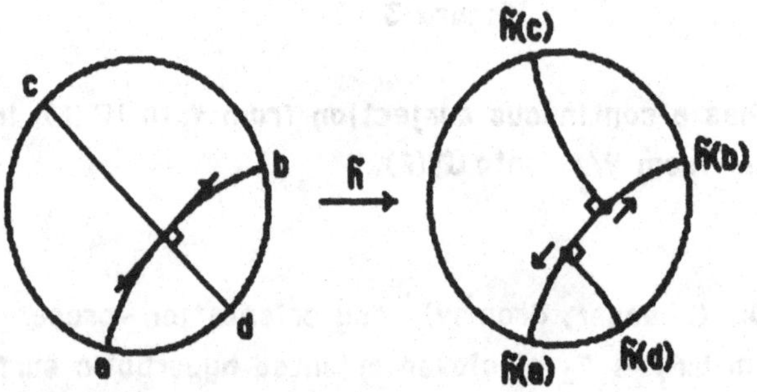

Figure 3.13

Remarks:

1) In Figure 3.13 both points (a,b,c), (b,a,d) $\in$ Y project to the same point of PT(F), but $\bar{h}$(a,b,c) and $\bar{h}$(b,a,d) project to different points. This shows that Theorem 3.10 does not work for PT(F).

2) If $\hat{\gamma}_+, \hat{\gamma}_- \subset UT(F_1)$ are the two lifts of an unoriented geodesic γ in F_1, then $\hat{h}(\hat{\gamma}_+)$, $\hat{h}(\hat{\gamma}_-)$ are the two lifts of an unoriented geodesic $\hat{h}(\gamma)$ in F_2. A more direct description of $\hat{h}(\gamma)$ is given in Figure 3.14.

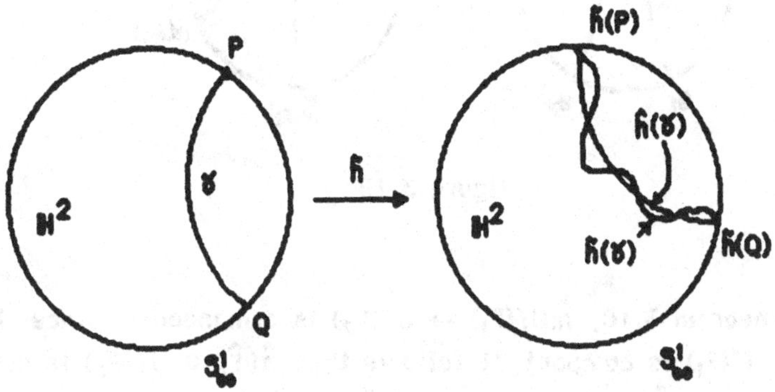

Figure 3.14

Theorem 3.11: Any orientation-preserving homeomorphism h: $F_1 \to F_2$ of closed oriented hyperbolic surfaces induces a homeomorphism $\hat{h}:\Lambda(F_1) \to \Lambda(F_2)$. If h is homotopic to k then $\hat{h} = \hat{k}$; if $h_1: F_1 \to F_2$ and $h_2: F_2 \to F_3$ are homeomorphisms then $\widehat{h_2 h_1} = \hat{h}_2 \cdot \hat{h}_1$.

Proof: For each lamination $L \in \Lambda(F_1)$ let $\hat{h}(L) = \bigcup_{\gamma \in L} \hat{h}(\gamma)$.

Let γ_1 and γ_2 be leaves of L with endpoints a_1, b_1 and a_2, b_2 respectively. Since $\tilde{h}$ preserves order on S^1_∞, and a_1, b_1 do not separate a_2, b_2, it follows that $\tilde{h}(a_1)$, $\tilde{h}(b_1)$ do not separate $\tilde{h}(a_2)$, $\tilde{h}(b_2)$. Therefore the leaves of $\hat{h}(L)$ are disjoint and simple.

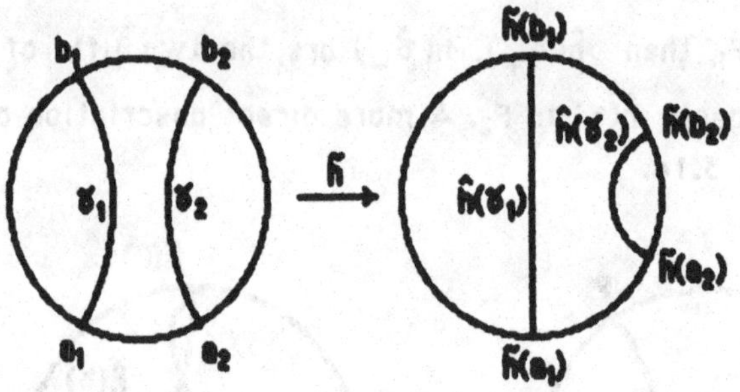

Figure 3.15

By Theorem 3.10, $\hat{h}: UT(F_1) \to UT(F_2)$ is continuous. Since the lift $\hat{L}$ of L to $UT(F_1)$ is compact, it follows that $\hat{h}(\hat{L}) \subset UT(F_2)$ is compact. This shows that $\hat{h}(L) \subset F_2$ is compact and hence a lamination.

To see that $\hat{h}$ is continuous, notice that the map $\hat{h}_*$ in the diagram below is continuous.

$$
\begin{array}{ccccccc}
\hat{\Lambda}_1 & \subset & 2^{UT(F_1)} & \xrightarrow{\hat{h}_*} & 2^{UT(F_2)} & & \\
{\scriptstyle p_{1*}}\downarrow{\scriptstyle \cong} & & \downarrow{\scriptstyle p_{1*}} & & {\scriptstyle p_{2*}}\downarrow & & \\
\Lambda_1 & \subset & 2^{F_1} & & 2^{F_2} & \supset & \Lambda_2
\end{array}
\qquad \Lambda_1 \xrightarrow{\hat{h}} \Lambda_2
$$

It follows that $\hat{h}\cdot p_{1*}|\hat{\Lambda}_1 = p_{2*}\cdot \hat{h}_*$ is continuous, so $\hat{h}$ is continuous.

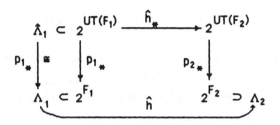

§4. Structure of Geodesic Laminations

In this chapter we show that geodesic laminations have rather few sublaminations. Let L be a lamination in a closed hyperbolic surface F. As usual, the universal cover $\tilde{F}$ of F is identified with $\mathbb{H}^2$.

Definition: A component of F − L is a *principal region* for L.

Lemma 4.1: Let U be a principal region for L and let $\tilde{U}$ be a component of the preimage of U in $\mathbb{H}^2$. Then $\overline{\tilde{U}}$ is a contractible hyperbolic surface with geodesic boundary.

Proof: The surface $\tilde{U}$ is a component of $\mathbb{H}^2 - \tilde{L}$ where $\tilde{L}$ is the preimage of L in $\mathbb{H}^2$. Distinct points a,b of $\tilde{U}$ cannot be separated by any leaf of $\tilde{L}$. It follows that the geodesic segment joining a and b is contained in $\tilde{U}$; in other words $\tilde{U}$ is *hyperbolically convex.*

If $\tilde{x} \in Fr(\tilde{U})$, then $\tilde{x}$ is on some leaf $\tilde{\ell}$ of $\tilde{L}$. If a is a point in $\tilde{U}$, and N is the hyperbolically convex hull of a and $\tilde{\ell}$, then $N - \tilde{\ell} \subset \tilde{U}$. Hence $\tilde{\ell} \subset \overline{\tilde{U}}$ (see Figure 4.1 below). It follows that $\overline{\tilde{U}}$ is a contractible hyperbolic surface with geodesic boundary.

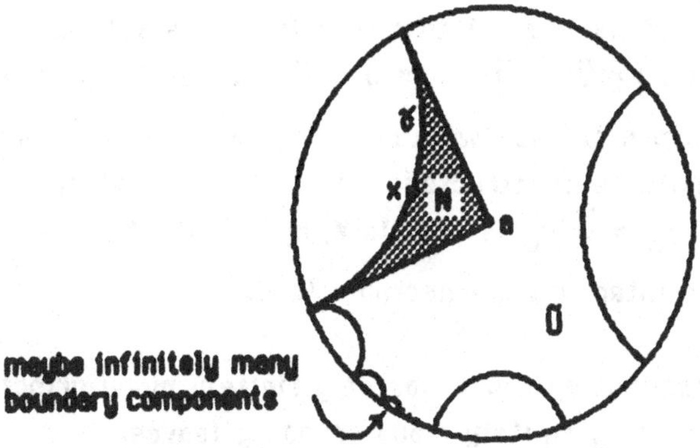

Figure 4.1

Exercise: If U is a principal region for $L \subset F$, then $\pi_1(U) \to \pi_1(F)$ is injective.

Definition: A *boundary leaf* of a principal region U is a leaf γ of L such that for all x in γ there exists an $\epsilon > 0$ such that $N_\epsilon(x) \cap U$ contains at least one component of $N_\epsilon(x) - \sigma$, where σ is the length 2ϵ segment of γ centered at x.

Remark: A boundary leaf is isolated from one side. The frontier of Ũ is contained in the preimage of the union of the boundary leaves of U.

Lemma 4.2: The union of all the boundary leaves of L is dense in L.

Proof: If $x \in L$, then arbitrarily close to x there exists a point u in $F - L$, as L is nowhere dense. Let v be the first point on the geodesic arc from u to x which is in L, then v is on a boundary leaf of L.

It follows from Lemma 4.1 that $\tilde{U}$ is the universal cover of U. If we set $\Gamma_U = \{g \in \Gamma \mid g(\tilde{U}) = \tilde{U}\}$, then $U = \tilde{U}/\Gamma_U$. Although the closure of U in F is seldom a surface with boundary having U as interior, such a surface can easily be constructed. For each principal region U, the quotient surface $V_U = \overline{\tilde{U}}/\Gamma_U$ is complete, hyperbolic, has geodesic boundary, and has interior homeomorphic to U.

Lemma 4.3: A lamination $L \subset F$ has only finitely many principal regions, each with only finitely many boundary leaves.

Proof: Let G be the disjoint union of the surfaces V_U for all the principal regions U for L. Note that

$$\text{Area}(G) = \text{Area}(G-\partial G) = \text{Area}(F-L) \leq \text{Area}(F).$$

Since G is a complete hyperbolic surface with finite area and geodesic boundary, Theorem 2.10 implies that G has finitely many components, each with finitely many boundary components.

Remarks:

1) Boundary leaves correspond to the boundary components of G, but this correspondence is not always 1-1.

2) The area of a principal region is $n\pi$ for some $n > 0$, therefore there are at most $-2\chi_F = 4g-4$ principal regions.

Exercise: The total number of boundary leaves is at most $-6\chi_F = 12g-12$.

Definition: A *crown* is a complete hyperbolic surface W with finite area and geodesic boundary, which is homeomorphic to $(S^1 \times [0,1]) - A$ for some finite subset A of $S^1 \times \{1\}$. Note that ∂W has exactly one circle component.

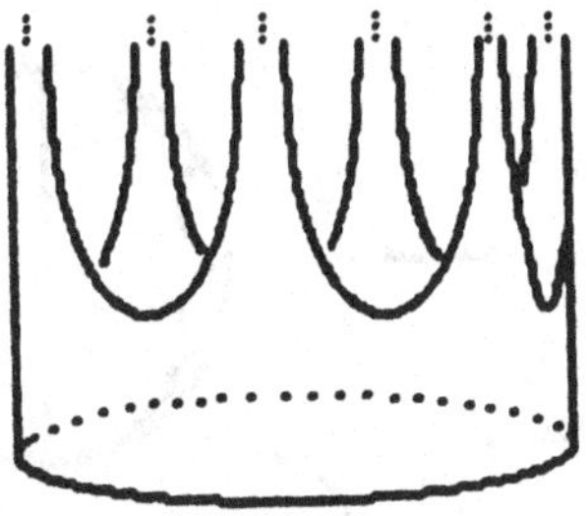

A crown

Figure 4.2

Lemma 4.4: Let L be a geodesic lamination without closed leaves on a closed orientable hyperbolic surface F. If U is a principal region, then either U is isometric to a finite sided polygon with vertices on S^1_∞, or there is a unique compact subset U_0 of U such that $U - U_0$ is isometric to a finite disjoint union of interiors of crowns.

Definition: The set U_0 is the *core* of U and each component of $U - U_0$ is a *crown* of U.

Examples:

1) The surface U obtained from a finite sided polygon in $\mathbb{H}^2$ as

indicated in Figure 4.3 is homeomorphic to a closed annulus with a point removed from each boundary component. The universal cover $\tilde{U}$ is obtained by replicating the pattern via the "gluing" isometry along its axis. This axis is mapped to the core of U.

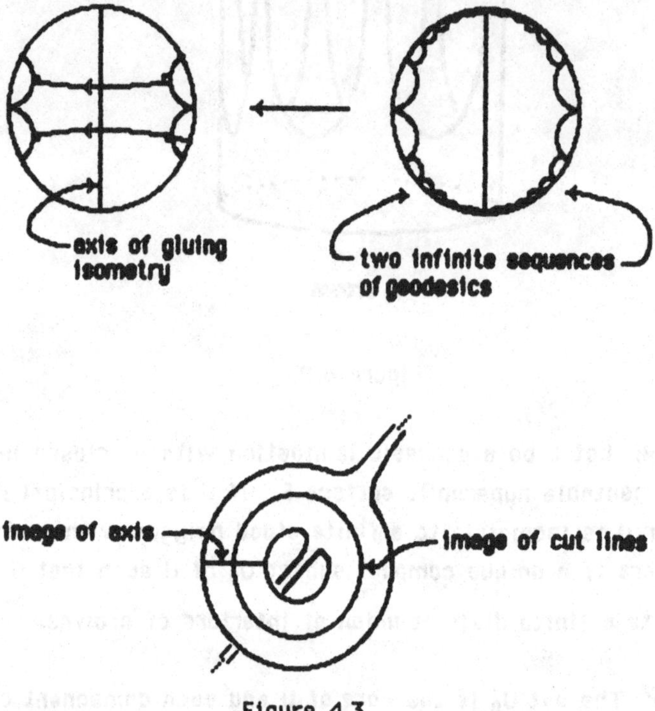

Figure 4.3

2) The surface U formed by gluing two hyperbolic polygons as illustrated in Figure 4.4 is homeomorphic to a pair of pants (see Figure 2.15) with two points removed from each boundary component. Here U

has a compact core consisting of a hyperbolic pair of pants, and three crowns.

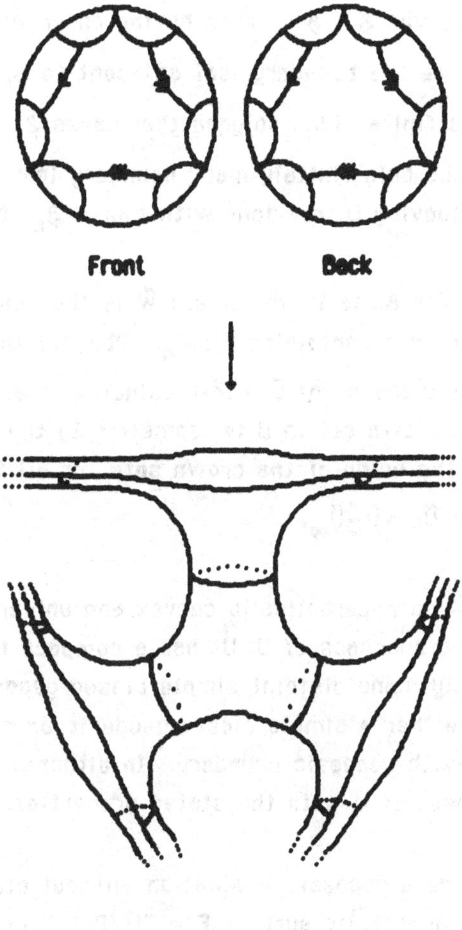

Figure 4.4

Proof of Lemma 4.4: Let $\tilde{U} \subset \mathbb{H}^2$ be a component of the preimage of U under the universal covering map. By hypothesis, no boundary leaf of U is closed. By Theorem 2.10, every boundary leaf β_0 of $\tilde{U}$ has two adjacent boundary leaves β_1, β_{-1} each having an endpoint in common with β_0. Let $\beta_{n\pm1}$ be the boundary leaf adjacent to β_n other than $\beta_{n\mp1}$; if U is not a finite sided polygon the leaves β_n $(n \in \mathbb{Z})$ are all distinct. Since U has only finitely many boundary leaves, there is a deck translation g leaving U invariant with $g\beta_0 = \beta_n$ for some $n > 0$.

The *crown set* for β_0 is $\tilde{W}-\partial\tilde{W}$ where $\tilde{W}$ is the smallest closed hyperbolically convex set containing all β_n. Observe that the crown sets for two boundary leaves of $\tilde{U}$ either coincide or are disjoint, and that the image of a crown set in U is isometric to the interior of a crown. Let $\tilde{U}_\infty$ be the union of the crown sets for all the boundary leaves of $\tilde{U}$, and let $\tilde{U}_0 = \tilde{U}-\tilde{U}_\infty$.

Observe that $\tilde{U}_0$ is hyperbolically convex and universally covers its image U_0 in U. As a subspace of U, U_0 has a compact frontier consisting of finitely many disjoint simple closed geodesics. It follows that U_0 is either a simple closed geodesic or a compact connected surface with geodesic boundary. In either case, U_0 is the unique compact subset of U with the stated properties.

Lemma 4.5: Let L be a geodesic lamination without closed leaves on a closed orientable hyperbolic surface $F = \mathbb{H}^2/\Gamma$. Then no point on S^1_∞ is an endpoint of infinitely many leaves of $\tilde{L}$, the preimage of L in $\mathbb{H}^2$.

Proof: Suppose $x \in S^1_\infty$ is an endpoint of infinitely many leaves of $\tilde{L}$. Boundary leaves are dense in $\tilde{L}$, so x is the endpoint of infinitely many boundary leaves of $\tilde{L}$. Since L has only finitely many oriented boundary leaves, there is a translation $g \in \Gamma - \{1\}$ fixing x. Let $\tilde{C}$ be the axis of g, and let $\tilde{\gamma}$ be a leaf of $\tilde{L}$ with endpoint x. The geodesic $\tilde{C}$ is in the closure of $\bigcup_{n \in \mathbb{Z}} g^n(\tilde{\gamma})$ and therefore in $\tilde{L}$. It follows that $\tilde{C}$ covers a closed leaf of L.

Lemma 4.6: Any closed leaf C in a geodesic lamination L on a closed orientable hyperbolic surface has a neighborhood N such that $L' \cap N \subseteq C$.

Proof: As usual let $F = \mathbb{H}^2/\Gamma$, let $\tilde{L}$ be the preimage of L and let $\tilde{C}$ be a component of the preimage of C. Let g be the element of Γ represented by C with axis $\tilde{C}$, moving points on $\tilde{C}$ a distance d, say. There is an $\epsilon > 0$ such that if $\tilde{\gamma} \subset \mathbb{H}^2$ is a geodesic with $0 < d(\tilde{C},\tilde{\gamma}) < \epsilon$, then the distance between the feet of the perpendiculars from the endpoints of $\tilde{\gamma}$ onto $\tilde{C}$ exceeds d, see Figure 4.5.

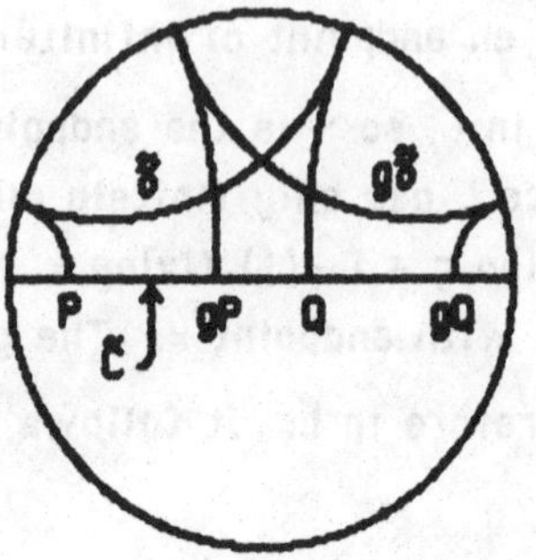

Figure 4.5

This forces $g\tilde{\gamma}$ to intersect $\tilde{\gamma}$ transversely. Let $\tilde{N}$ be the ϵ-neighborhood of $\tilde{C}$, and let N be the image of $\tilde{N}$ in F. It follows that any leaf $\tilde{\gamma}$ of $\tilde{L}$ which meets $\tilde{N}$ has an endpoint in common with $\tilde{C}$. The argument of Lemma 4.5 shows that there are only finitely many leaves of $\tilde{L}$ between $\tilde{\gamma}$ and $g\tilde{\gamma}$, so all leaves meeting $\tilde{N}-\tilde{C}$ are isolated.

Theorem 4.7: If L is a geodesic lamination on a closed orientable hyperbolic surface F and L_1 is a sublamination of L, then $L_1 \cap L'$ is a union of components of L'.

Corollary 4.7.1: For any lamination L, L''' = L''; if L has no closed leaves, then L'' = L'.

Proof of 4.7.1: By Lemma 4.2, L' has finitely many components $K_1, \ldots, K_r$, so $L'' = K_1' \cup \ldots \cup K_r'$. Applying 4.7 to $L'' \subseteq L$ and after renumbering, $L'' = K_1 \cup \ldots \cup K_s$ $(s \leq r)$. So $K_i' = K_i$ $(i \leq s)$ giving $L''' = K_1' \cup \ldots \cup K_s' = K_1 \cup \ldots \cup K_s = L''$. If $L'' \neq L'$, then $s < r$, so K_r' is empty and so not equal to K_r. By Lemma 3.6, K_r consists of closed leaves, so L has a closed leaf.

Corollary 4.7.2: Every leaf of L is dense in L if and only if L is connected and L' = L.

Proof of 4.7.2: Let γ be a leaf of L and let $L_1 = \overline{\gamma} \subseteq L$. If L is connected and L' = L then by Theorem 4.7, $L_1 = L$, so γ is dense in L. The converse is clear.

Proof of Theorem 4.7: Let L_2 be the union of all non-closed leaves of $L_1 \cap L'$. By Lemma 4.6, L_2 is closed in F, and is therefore a lamination without closed leaves. We shall prove that the intersection of L' with any principal region U for L_2 is contained in the core U_0 of U. This will imply that $L' - L_2$ is equal to $L' \cap V_0$, where V_0 is the union of the cores of all the principal regions for L_2. As V_0 is compact, it will follow that L_2 is open and closed in L', and that $L_1 \cap L' = L_2 \cup$ (isolated leaves) is open and closed in L'.

As usual, identify $\tilde{F}$ with $\mathbb{H}^2$, let $\tilde{L}$ be the preimage of L, $\tilde{L}'$ the preimage of L', and $\tilde{U}$ a component of the preimage of U. Observe that any leaf of $\tilde{L}$ meeting $\tilde{U}$ is contained in $\tilde{U}$. If $\tilde{U}$ is a finite sided polygon, then $\tilde{U} \cap \tilde{L}$ consists of finitely many diagonals of $\tilde{U}$, all isolated, so $\tilde{U} \cap \tilde{L}'$ is empty. Otherwise, by Lemma 4.4, U has a core U_0 with covering $\tilde{U}_0 \cap \tilde{U}$ such that each component of $\tilde{U} - \tilde{U}_0$ is a universal cover of a crown. Any leaf of $\tilde{U} \cap \tilde{L}$ not contained in $\tilde{U}_0$ has one end in such a component W. It follows from Lemma 4.5 that each such leaf is isolated, so $\tilde{U} \cap \tilde{L}' \subset \tilde{U}_0$ as required.

Lemma 4.8: Let F_1, F_2 be closed orientable hyperbolic surfaces, let $h:F_1 \to F_2$ be a homeomorphism, and let L be a geodesic lamination of F_1. Then $(\hat{h}(L))' = \hat{h}(L')$ and for any principal region U for L there is a unique principal region $\hat{h}(U)$ for $\hat{h}(L)$ whose boundary leaves are $\hat{h}$(boundary leaves of L). If U has core U_0 then $\hat{h}(U)$ has core $\hat{h}(U_0)$ with frontier $\hat{h}$(frontier of U_0).

Proof: The notions of isolated leaf, boundary leaf, and core are all definable in terms of the cyclic ordering on S^1_∞, which is preserved by $\hat{h}$. Details are left to the reader.

Theorem 4.9: A geodesic lamination L on a closed orientable hyperbolic surface F has measure 0. Equivalently, the area of F − L is equal to $-2\pi\chi_F$, the area of F.

Remark: We only sketch the proof; for more details see [FLP].

Proof:

Step 1. There exists a continuous, integrable line field on F, tangent to L everywhere, with only finitely many isolated singularities of the types illustrated in Figure 4.6.

Recall: A line field is locally a non-oriented version of a nowhere-zero vector field.

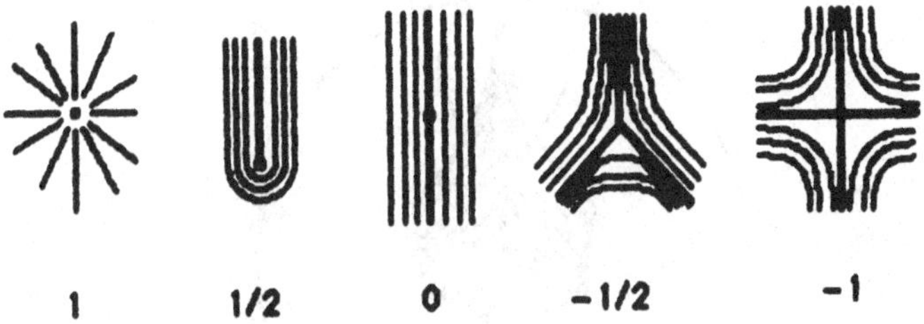

Figure 4.6

This line field is constructed first on principal regions. Indeed, on the fundamental domain of such, the line field is an extension of the tangent field to the lamination.

Example: Figure 4.7 illustrates a fundamental domain for a principal region of genus 1 with one boundary leaf. (The angles appearing in Figure 4.7 should have sum π to ensure that the boundary is geodesic.) Begin the line field in a neighborhood of an ideal point by choosing the geodesics with the ideal point as endpoint. The line field will be parallel to the unidentified boundary components of the fundamental region. What is left of the region is compact, hence the line field may be extended continuously by coning the ideal vertices and centers of finite edges to form the separatrices and extending the line field as illustrated in Figure 4.7.

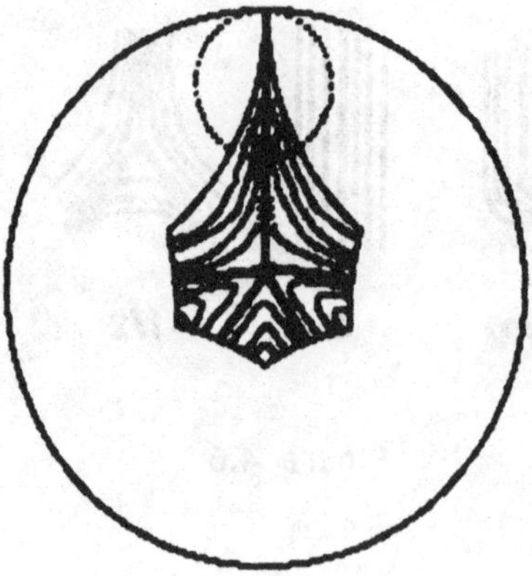

Figure 4.7

Definition: The *index* of a singularity is the winding number of the line field restricted to the boundary of a disk neighborhood of the singularity. This can be computed by orienting the line field locally and counting the number of rotations of a tangent vector as one goes around the boundary of a disk neighborhood counter-clockwise (see Figure 4.6).

Step 2. The sum of the indices of all the singularities is equal to χ_F.

Regard F as a glued polygon with vertices at regular points and edges transverse to the line field.

Example: A genus 2 surface:

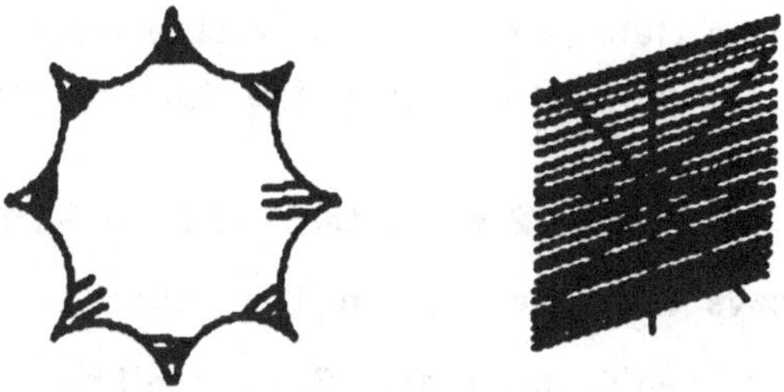

Figure 4.8

The line field points to two interior angles and cuts across all the others. The total index is the winding number around the perimeter of the polygon, which is equal to χ_F.

Step 3. As in Lemma 4.3, $F - L$ is isometric to $G - \partial G$ where G is a complete hyperbolic surface with geodesic boundary and finite area. The surface G inherits a line field which is tangent to ∂G and as in Figure 4.9 near the ends of G.

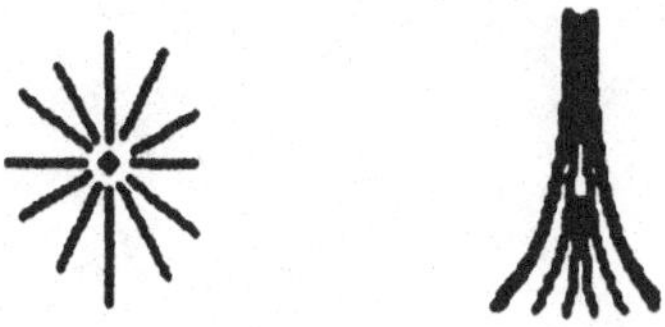

(a) (b)

Figure 4.9

We generalize step 2 to show that the sum of the indices at the singularities is equal to $\chi_G - (1/2)\chi_{\partial G}$. If G is unbounded this formula holds provided the line field is as shown in Figure 4.9(a) near the ends of G. The general case now follows from the doubling trick.

Step 4. It follows from steps 2 and 3 that $-2\chi_F = -2\pi\chi_G + \pi\chi_{\partial G}$. Therefore F and G have the same area, implying that

$$\text{area}(F-L) = \text{area}(G-\partial G) = \text{area}(G) = \text{area}(F).$$

§5. Surface Automorphisms

Now we are ready to give a version of Nielsen's classification of surface automorphisms.

Definition: An automorphism $h:F \to F$ of a closed orientable surface is called *periodic* if h^n is homotopic to the identity for some $n > 0$. An automorphism is *reducible* if h is homotopic to an automorphism which leaves invariant some essential closed 1-submanifold of F.

Remarks:

1) It might seem more natural to call h periodic only if h^n is the identity for some $n > 0$. In fact, if h^n is homotopic to the identity, then h is isotopic to an automorphism g such that g^n is the identity. This was first proved by Fenchel and Nielsen; see the remarks following Lemma 3.8.

2) Assuming F has a hyperbolic structure, an automorphism h is reducible if and only if $\hat{h}(C) = C$ for some geodesic 1-submanifold C of F.

In this chapter we study automorphisms which are neither periodic nor reducible; first we give examples of such automorphisms.

Lemma 5.1: Let $h:F \to F$ induce $h_*: H_1(F) \to H_1(F)$, having matrix A with respect to some basis. If the characteristic polynomial $\chi_h(t)$ of A is irreducible over $\mathbb{Z}$, has no roots of unity as zeros, and is not a polynomial in t^n for any $n > 1$, then h is irreducible and non-periodic.

Proof: If h is periodic, then $(h_*)^n = 1$ for some $n > 0$, implying that all the zeros of $\chi_h(t)$ are n-th roots of unity.

If h is reducible, then after homotopy $h(C) = C$ for some 1-submanifold C whose components may be assumed to be essential geodesics. There are two cases.

Case 1. Some component C_1 of C does not separate F.
For some $n > 0$, $(h_*)^n[C_1] = [C_1] \neq 0$, so $(h_*)^n$ has eigenvalue 1. It follows that h_* has an n-th root of unity as an eigenvalue.

Case 2. All components of C separate F.
There is a component F_0 of F − C having as frontier a single component of C.

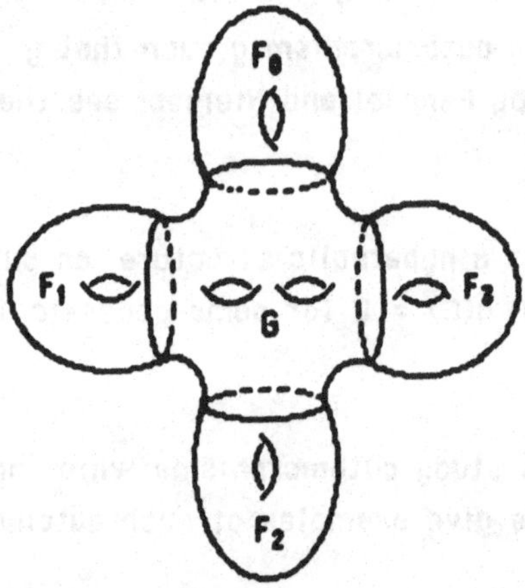

Figure 5.1

If $F_r = h^r(F_0)$, there exists a least n such that $F_n = F_0$. Note that

$$H_1(F) \cong H_1(F_0) \oplus \ldots \oplus H_1(F_{n-1}) \oplus H_1(\overline{G}),$$

where $\overline{G}$ is obtained by capping off the boundary components of G (see Figure 5.1).

Since h_* permutes the $H_1(F_i)$ cyclically, there is a basis for $H_1(F)$ so that the matrix A of h_* is of the form illustrated below (for n = 4).

$$A = \begin{bmatrix} 0 & 0 & 0 & B & 0 \\ I & 0 & 0 & 0 & 0 \\ 0 & I & 0 & 0 & 0 \\ 0 & 0 & I & 0 & 0 \\ 0 & 0 & 0 & 0 & C \end{bmatrix}$$

Hence $\chi_h(t) = |A - tI| = |B - t^nI| \cdot |C - tI|$. As $\chi_h(t)$ is irreducible and F_0 has genus at least 1, it follows that $\chi_h(t) = |B - t^nI|$ (and G is a punctured sphere).

Example: Let T_i be the Dehn twist in the curve C_i as illustrated in Figure 5.2 (positive twist to the right).

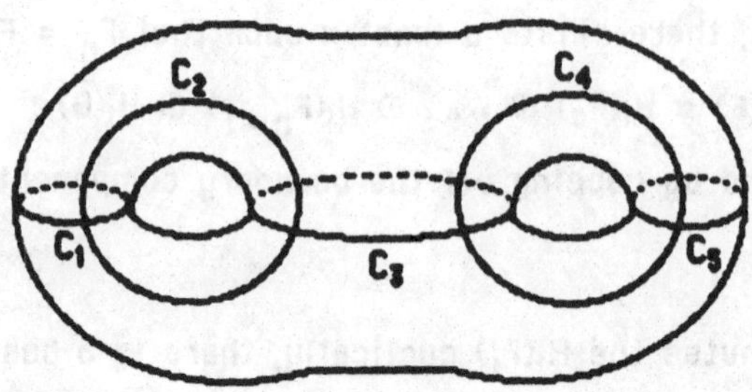

Figure 5.2

Claim: $T_1T_3(T_5)^2(T_2)^{-1}(T_4)^{-1}$ is irreducible and non-periodic.

Remark: Penner shows that more is true. The automorphism given by any finite sequence of the T_i's where the odd indices carry the same sign and even indices carry the opposite sign is irreducible and non-periodic.

Proof of claim: Using the basis:

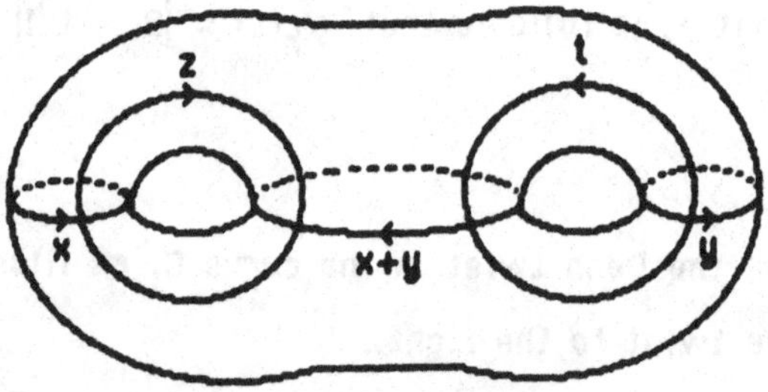

Figure 5.3

the matrix for each Dehn twist is

$$T_1^* = \begin{bmatrix} 1 & 0 & 1 & 0 \\ 0 & 1 & 0 & 0 \\ 0 & 0 & 1 & 0 \\ 0 & 0 & 0 & 1 \end{bmatrix}, \quad T_3^* = \begin{bmatrix} 1 & 0 & 1 & 1 \\ 0 & 1 & 1 & 1 \\ 0 & 0 & 1 & 0 \\ 0 & 0 & 0 & 0 \end{bmatrix}, \quad T_5^* = \begin{bmatrix} 1 & 0 & 0 & 0 \\ 0 & 1 & 0 & 1 \\ 0 & 0 & 1 & 0 \\ 0 & 0 & 0 & 1 \end{bmatrix}$$

$$T_2^* = \begin{bmatrix} 1 & 0 & 0 & 0 \\ 0 & 1 & 0 & 0 \\ 1 & 0 & 1 & 0 \\ 0 & 0 & 0 & 1 \end{bmatrix}, \quad T_4^* = \begin{bmatrix} 1 & 0 & 0 & 0 \\ 0 & 1 & 0 & 0 \\ 0 & 0 & 1 & 0 \\ 0 & 1 & 0 & 1 \end{bmatrix}$$

Exercise: The characteristic polynomial of $T_1T_3(T_5)^2(T_2)^{-1}(T_4)^{-1}$ is $t^4 - 9t^3 + 21t^2 - 9t + 1$, is irreducible over $\mathbb{Z}$, and has no roots of unity as zeros.

Definition: If L_1 and L_2 are laminations then $L_1 \pitchfork L_2$ denotes the set of transverse intersection points of L_1 with L_2.

Lemma 5.2: If $h:F \to F$ is a non-periodic automorphism of a closed orientable hyperbolic surface, then $\hat{h}(L) = L$ for some $L \in \Lambda(F)$.

Corollary 5.2.1: There is a lift $\tilde{k}$ of a strictly positive power of h such that the restriction of $\tilde{k}$ to S^1_∞ has at least two fixed points.

Proof of 5.2.1: By Lemmas 4.3 and 4.8, $\hat{h}$ permutes the finite set of oriented boundary leaves of L. Choose $m > 0$ so that $\hat{h}^m$ maps an oriented boundary leaf γ to itself. Let $\tilde{\gamma}$ be a component of the preimage of γ and let $\tilde{j}$ be any lift of h^m. Then $\tilde{j}(\tilde{\gamma}) = g\tilde{\gamma}$ for some covering translation g. Then $\tilde{k} = g^{-1}\tilde{j}$ is also a lift of h^m and fixes both endpoints of $\tilde{\gamma}$.

Proof of 5.2: By Theorem 2.7 there is a simple closed geodesic $C \subset F$ such that $\hat{h}^n(C) \neq C$ for all $n > 0$. It follows that $\hat{h}^m(C) \neq \hat{h}^n(C)$ if $m \neq n$.

Since $\Lambda(F)$ is compact, the sequence $\hat{h}^n(C)$ has a convergent subsequence, say $\hat{h}^{n_i}(C)$ converges to $K \in \Lambda(F)$. Since the geodesics $\hat{h}^{n_i}(C)$ are distinct, K is not an isolated point of $\Lambda(F)$; by Lemma 3.6, K' is non-empty.

For a fixed integer r, Theorem 3.11 implies that $\hat{h}^{n_i+r}(C)$ converges to $\hat{h}^r(K)$. Note that $|\hat{h}^{n_i}(C) \cap \hat{h}^{n_i+r}(C)| = |C \cap \hat{h}^r(C)|$. Let $N_r = |C \cap \hat{h}^r(C)|$ and suppose that $|K \pitchfork \hat{h}^r(K)| > N_r$. Choose disjoint neighborhoods of $N_r + 1$ points in $K \pitchfork \hat{h}^r(K)$; if n_i is sufficiently large, each contains at least one point of $\hat{h}^{n_i}(C) \cap \hat{h}^{n_i+r}(C)$. This is impossible, so $|K \pitchfork \hat{h}^r(K)| \leq N_r$ is finite. It follows that $K \pitchfork \hat{h}^r(K')$ is empty.

For any integers r and s, $\hat{h}^r(K')$ has no transverse intersections with $\hat{h}^s(K')$. So the union E of all the sets $\hat{h}^r(K')$ is a non-empty disjoint union of simple geodesics. By Lemma 3.2, $L = \bar{E}$ is a lamination; clearly $\hat{h}(L) = L$. Observe (for future use) that $K \pitchfork L$ is empty.

Lemma 5.3: If $h:F \to F$ is an irreducible automorphism of a closed orientable hyperbolic surface and $\hat{h}(L) = L$ for some $L \in \Lambda(F)$, then each component of $F - L'$ is contractible (and hence a finite sided polygon) and each leaf of L is dense in L'.

Proof: The union of all closed leaves of L is a geodesic 1-submanifold invariant under $\hat{h}$. Since h is irreducible, L has no closed leaves. By Lemma 4.4, any non-contractible component U of $F - L$ has a non-empty core U_0. If V_0 is the union of the cores of all the components of $F - L$, then $Fr(V_0)$ is a 1-submanifold invariant under $\hat{h}$. Therefore V_0 is empty, and all the components of $F - L$ are contractible.

Since $\hat{h}(L') = L'$, all components of $F - L'$ are contractible, implying that L' is connected. If γ is a leaf of L and $L_1 = \overline{\gamma}$, then $L_1 \cap L' \supseteq L_1'$ which is non-empty by Lemma 3.6. By Theorem 4.7, γ is dense in L'.

Definition: Let L be a geodesic lamination on an orientable hyperbolic surface F, and let $\tilde{L}$ be the preimage of L in the universal covering of F. A *stable interval (for* L*)* is a closed interval $I \subset S^1_\infty$ such that for any two points $P,Q \in \mathrm{Int}I$ there is a leaf $\tilde{\gamma} \subset \tilde{L}$ whose endpoints separate P and Q from ∂I.

Lemma 5.4: If $h:F \to F$ is a non-periodic irreducible automorphism of a closed oriented hyperbolic surface, then $\hat{h}(L) = L$ for some $L \in \Lambda(F)$ with the following property. If a lift $\tilde{k}$ of a strictly positive power of h maps a stable interval I onto itself, then the restriction of $\tilde{k}$ to I has a fixed point $Z \in \mathrm{Int}I$ such that for all $P \in I - \{Z\}$, $\tilde{k}^n(P)$ converges to a point in ∂I.

Remark: Since $\tilde{k}$ is orientation-preserving and $\tilde{k}(I) = I$, the endpoints of I are fixed by $\tilde{k}$. The endpoints are *contracting* fixed points of the restriction of $\tilde{k}$ to I and Z is an *expanding* fixed point; note that the restriction of $\tilde{k}$ to I has no other fixed points.

Proof: Let L be the lamination constructed in Lemma 5.2. Recall from that construction that $\hat{h}(L) = L$ and $K \pitchfork L$ is empty, where K is the limit of $\hat{h}^{n_i}(C)$ for some simple closed geodesic C and some sequence $n_i \to \infty$. Suppose that $\tilde{k}$ is a lift of h^m (where $m > 0$) and I is a stable interval with $\tilde{k}(I) = I$. It follows easily from the definition that the geodesic $\tilde{\gamma}$ joining the endpoints of I is a leaf of $\tilde{L}$. By Lemma 5.3, the image γ of $\tilde{\gamma}$ is dense in L′ and $C \cap L'$ is non-empty, so $C \cap \gamma$ is also non-empty. Therefore, some component $\tilde{C}$ of the preimage of C has endpoints $A \in \mathrm{Int}I$, $B \notin I$. Clearly $\tilde{k}^n(A)$, $\tilde{k}^n(B)$ converge to A_∞, B_∞, with $A_\infty \in I$ and $B_\infty \notin \mathrm{Int}I$.

If $A_\infty \in \mathrm{Int}I$, there is a leaf $\tilde{\delta} \subset \tilde{L}$ whose endpoints separate A_∞ from ∂I. Recall that $\hat{h}^{n_i}(C)$ converges to K and that $\tilde{k}$ is a lift of h^m. There must be infinitely many n_i in some residue class modulo m, so $\hat{h}^{mq_i}(C)$ must converge to $\hat{h}^r(K)$ for some sequence $q_i \to \infty$ and some

$r \in \mathbb{Z}$. The geodesic $\tilde{C}_\infty$ with endpoints A_∞, B_∞ is a lift of a leaf of $\hat{h}^r(K)$ which meets the leaf $\tilde{\delta}$ of $\tilde{C}$ transversely. But K, and hence $\hat{h}^r(K)$, has no transverse intersection with L. This contradiction shows that $A_\infty \in \partial I$. Observe that A_∞ is a contracting fixed point of the restriction of $\tilde{k}$ to I.

Let U be the open subinterval of I with endpoints A, A_∞; then $\tilde{k}$ moves all points of U strictly closer to A_∞. Since $\tilde{k}$ is continuous and restricts to the identity on ∂I, there is a neighborhood V of $\partial I - \{A_\infty\}$ such that V and $\tilde{k}(V)$ are disjoint from U. Since I is a stable interval there is a leaf $\tilde{\delta}$ of $\tilde{C}$ with endpoints $X \in U$, $Y \in V$. It follows that $\tilde{k}(X)$ and $\tilde{k}(Y)$ separate X and Y from ∂I. Let X_∞, Y_∞ be the limits of the sequences $\tilde{k}^{-n}(X)$, $\tilde{k}^{-n}(Y)$ (as as $n \to +\infty$).

If $X_\infty \neq Y_\infty$, then the closure of the component of $S^1_\infty - \{X_\infty, Y_\infty\}$ that is not contained in I is a stable interval J with $\tilde{k}(J) = J$. Both endpoints of J are expanding fixed points of the restriction of $\tilde{k}$ to J; but replacing I by J in the first part of the proof shows at least one endpoint must be contracting. Therefore, $X_\infty = Y_\infty$; it is easy to check that $Z = X_\infty = Y_\infty$ has the properties specified in the lemma.

Theorem 5.5: Let $h:F \to F$ be a non-periodic irreducible automorphism of a closed orientable hyperbolic surface. Any lift of a strictly positive power of h has finitely many fixed points on S^1_∞, alternately contracting and expanding (see Figure 5.4). There is a unique perfect lamination L^s, invariant under $\hat{h}$, such that $\tilde{L}^s$ contains the geodesics joining consecutive contracting fixed points of any lift of a strictly positive power of h. Every leaf of L^s is dense in L^s (see Figure 5.5).

Remark: L^s is the *stable lamination* of h. The *unstable lamination* L^u is the stable lamination of h^{-1}.

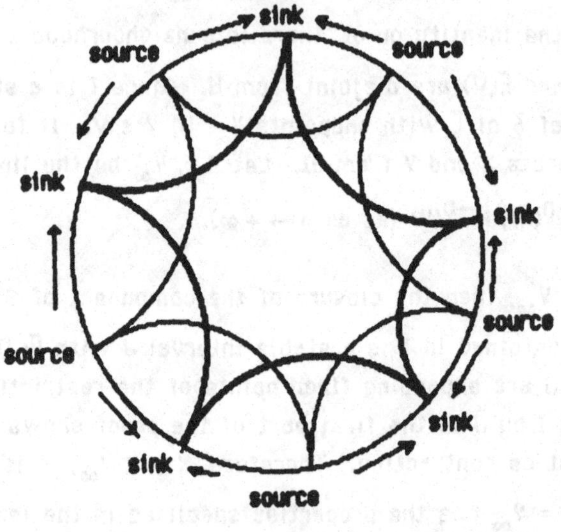

Figure 5.4

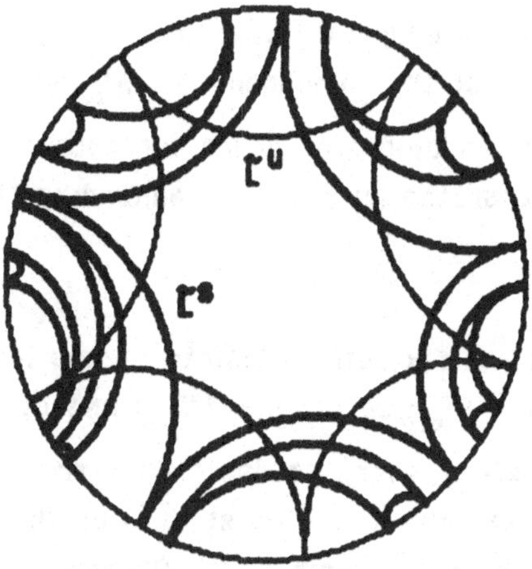

Figure 5.5

Proof: Let L be the lamination given by Lemma 5.4; by Lemma 5.3, L′ is perfect. Let $\tilde{k}$ be a lift of a strictly positive power of h; we consider three cases.

Case 1) $\tilde{k}$ fixes the endpoints of a boundary leaf $\tilde{\gamma}$ of $\tilde{L}'$. Since L′ is perfect, $\tilde{\gamma}$ is in the frontier of a unique component $\tilde{U}$ of $\mathbb{H}^2 - \tilde{L}'$. By Lemma 5.3, $\tilde{U}$ is a finite sided polygon, so $\tilde{k}$ fixes all the vertices of $\tilde{U}$.

It follows from Lemma 4.5 that the closures of the components of S^1_∞ − {vertices of $\tilde{U}$} are stable intervals. Therefore, the vertices of $\tilde{U}$ are contracting fixed points of $\tilde{k}$ restricted to S^1_∞, and between two consecutive vertices is one expanding fixed point (and no other fixed points).

Case 2) $\tilde{k}$ fixes the endpoints of a non-boundary leaf $\tilde{\gamma}$ of $\tilde{L}'$. In this case the closures of the components of S^1_∞ – {endpoints of $\tilde{\gamma}$} are stable intervals. The endpoints of $\tilde{\gamma}$ are contracting fixed points of $\tilde{k}$ restricted to S^1_∞, separated by a pair of expanding fixed points (and no other fixed points).

Case 3) $\tilde{k}$ does not fix both endpoints of any leaf of $\tilde{L}'$. Suppose $\tilde{k}$ fixes some point $x \in S^1_\infty$; by Lemma 4.5, x cannot be the endpoint of any leaf of $\tilde{L}'$. Therefore, every geodesic with endpoint x meets $\tilde{L}'$ transversely. For each leaf $\tilde{\gamma} \subset \tilde{L}'$, let $U(\tilde{\gamma})$ be the component of S^1_∞ – {endpoints of $\tilde{\gamma}$} not containing x. Observe that the sets $U(\tilde{\gamma})$ ($\tilde{\gamma} \subset \tilde{L}'$) cover S^1_∞ – {x}. If $\tilde{\gamma}_1, \tilde{\gamma}_2 \subset \tilde{L}'$ then $U(\tilde{\gamma}_1)$ and $U(\tilde{\gamma}_2)$ are either disjoint or nested. It follows that any compact subset of S^1_∞ – {x} is contained in $U(\tilde{\gamma})$ for some $\tilde{\gamma} \subset \tilde{L}'$.

There is a leaf $\tilde{\gamma}$ of $\tilde{L}'$ so that $U(\tilde{\gamma})$ has non-empty intersection with $\tilde{k}U(\tilde{\gamma})$; let A,B be the endpoints of $\tilde{\gamma}$. The sequences $\tilde{k}^n(A)$ and $\tilde{k}^n(B)$ have a common limit, which is a contracting fixed point of $\tilde{k}|S^1_\infty$. Similiarly, $\tilde{k}^{-n}(A)$ and $\tilde{k}^{-n}(B)$ converge to an expanding fixed point.

In all cases, $\tilde{k}$ restricted to S^1_∞ has finitely many fixed points, alternately contracting and expanding. Observe that L′ has all the properties claimed for L^s (the leaves of L′ are dense by Lemma 5.3).

Let $\tilde{\gamma}$ be a boundary leaf of $\tilde{L}'$. As in Corollary 5.2.1, there is a lift $\tilde{k}$ of h^m (for some $m > 0$) fixing the endpoints of $\tilde{\gamma}$. We showed above in case 1) that the endpoints of $\tilde{\gamma}$ are consecutive contracting fixed points of $\tilde{k}|S^1_\infty$. Since γ is dense in L', any lamination L^s with the properties stated coincides with L'.

Lemma 5.6: *If $h:F \to F$ is a non-periodic irreducible automorphism of a closed orientable hyperbolic surface and C is an essential closed curve in F, then $h^m(C)$ is homotopic to $h^n(C)$ if and only if $m = n$.*

Proof: We may suppose that C is a geodesic. If the conclusion to Lemma 5.6 fails for C, then K the union of all the geodesics $\hat{h}^q(C)$ (for $q \in \mathbb{Z}$) is actually a finite union of geodesics. It follows that the preimage of K is a closed subset of $\mathbb{H}^2$.

Let $\tilde{\gamma}$ be a lift of a leaf γ of L^u with endpoints X, Y in S^1_∞ which are fixed by a lift $\tilde{k}$ of a power of h. Since $C \pitchfork L^u$ is non-empty and γ is dense in L^u, there is a lift $\tilde{C}$ of C with $\tilde{C} \cap \tilde{\gamma}$ non-empty. Let A, B be the endpoints of $\tilde{C}$; as $n \to \infty$, the sequences $\tilde{k}^n(A)$, $\tilde{k}^n(B)$ converge to contracting fixed points A_∞, B_∞ of $\tilde{k}|S^1_\infty$. By Theorem 5.5, the geodesic $A_\infty B_\infty$ is either a leaf of $\tilde{L}^s$ or lies in a component of $\mathbb{H}^2 - \tilde{L}^s$. On the other hand, $A_\infty B_\infty$ is in the preimage of K, implying that some $\hat{h}^q(C)$ fails to meet L^s transversely, a contradiction.

Theorem 5.7: Let $h:F \to F$ be a non-periodic irreducible automorphism of a closed orientable hyperbolic surface. There is an integer $m > 0$ such that for any simple closed geodesic C in F, the sequence $\hat{h}^{mn}(C)$ converges to a lamination K_C as $n \to \infty$. The lamination K_C is one of the finitely many laminations containing L^s.

Proof: By Lemma 5.6, the homotopy classes $h^n(C)$ are all distinct, so are the closed geodesics $\hat{h}^n(C)$. Suppose that $\hat{h}^{n_i}(C) \to K \in \Lambda(F)$ as in Lemma 5.2: then $L^s = (\bigcup_{r \in \mathbb{Z}} \hat{h}^r(K'))'$ contains a leaf of K. Since each leaf is dense in L^s, $K \supseteq L^s$.

There are only finitely many laminations $K_1, \ldots, K_\ell$ containing L^s (as $K_i - L^s$ is a finite union of diagonals of components of $F - L^s$). There is an integer m (depending only on h, not on C) such that $\hat{h}^m$ fixes each K_i. It follows easily that $\hat{h}^{mn}(C)$ converges to one of $K_1, \ldots, K_\ell$ as $n \to \infty$.

§6. Pseudo-Anosov Automorphisms

The structure of non-periodic, irreducible automorphisms was greatly clarified by Thurston's introduction of "invariant measures" on the stable and unstable laminations. It is convenient to describe these measures on "singular foliations" associated with the laminations, rather than on the laminations themselves. One advantage to this approach is that each non-periodic, irreducible isotopy class is canonically represented by a "pseudo-Anosov" automorphism leaving the singular foliations invariant.

Lemma 6.1: Let $h: F \to F$ be a non-periodic irreducible automorphism of a closed oriented hyperbolic surface. Then h is isotopic to a homeomorphism $h': F \to F$ such that $h'(L^s) = L^s$ and $h'(L^u) = L^u$.
Remark: The homeomorphism h' is not necessarily a diffeomorphism.

Proof: Let $\tilde{L}^s$, $\tilde{L}^u$ be the preimages of L^s, L^u respectively in the universal cover of F and let $\tilde{h}$ be a lift of h extended in the usual way over $\mathbb{H}^2 \cup S^1_\infty$. The restriction of $\tilde{h}$ to S^1_∞ induces a bijection $\tilde{h}'$ on $\tilde{L}^s \cap \tilde{L}^u$, which is continuous because the angle of intersection between leaves of $\tilde{L}^s$ and $\tilde{L}^u$ is bounded away from 0 and π. This bijection can be extended over $\tilde{L}^s \cup \tilde{L}^u$ by extending linearly over each interval in $\tilde{L}^s - \tilde{L}^u$ or $\tilde{L}^u - \tilde{L}^s$ with respect to hyperbolic distance. One checks that $\tilde{h}': \tilde{L}^s \cup \tilde{L}^u \to \tilde{L}^s \cup \tilde{L}^u$ is continuous, hence uniformly so.

Notice that $F - (L^s \cup L^u)$ has regions whose closures are $2k$-gons. For $k \geq 3$, there is only one such region for each complementary region of L^s in F, hence only finitely many in all. On the other hand, there are infinitely many rectangles. We can extend $\tilde{h}'$ over the non-rectangles by any $\pi_1(F)$-equivariant choice of homeomorphism. To extend $\tilde{h}'$ over the rectangles, form a correspondence between opposite sides using hyperbolic distance, setting up a "grid" on the rectangle (see Figure 6.1).

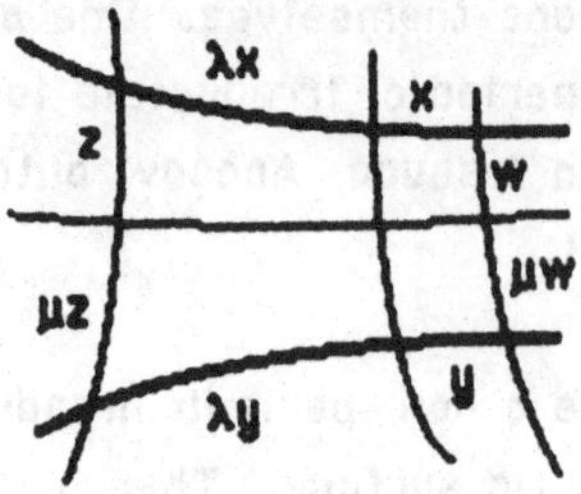

Figure 6.1

Extending $\tilde{h}'$ linearly over the grids gives a continuous extention $\tilde{h}':\mathbb{H}^2 \to \mathbb{H}^2$.

If $h':F \to F$ is the homeomorphism induced by $\tilde{h}'$, then $h'(L^s) = L^s$ and $h'(L^u) = L^u$. Since the lifts $\tilde{h}$ and $\tilde{h}'$ induce the same action on S^1_∞, h and h' are homotopic and therefore isotopic.

Definition: A *singular foliation* $\mathfrak{F}$ on a surface F is a decomposition of F as a disjoint union of *leaves.* Any point x in F outside of a finite set S has a chart φ: U $\to \mathbb{R}^2$ carrying the components of U $\cap$ leaf to horizontal intervals.

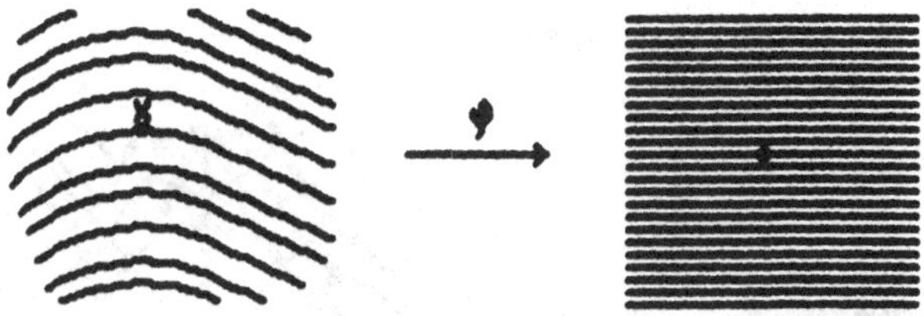

Figure 6.2

For x in S, x has a chart $\varphi: U \to \mathbb{R}^2$ taking $\mathcal{F} \cap U \to W_k$, where W_k is the standard "k-prong singularity" or "singularity with k separatrices" illustrated for k = 4 in Figure 6.3. We call S the *singular set* of $\mathcal{F}$.

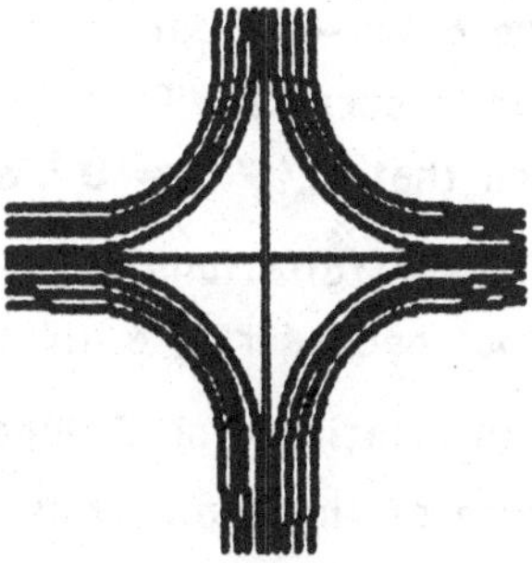

Figure 6.3

Two singular foliations $\mathcal{F}^s, \mathcal{F}^u$ are *transverse* if they have the same singular set and at all other points the leaves are transverse. Further, at the singular set we require the standard V_k model at singular points as illustrated in Figure 6.4 for $k = 4$.

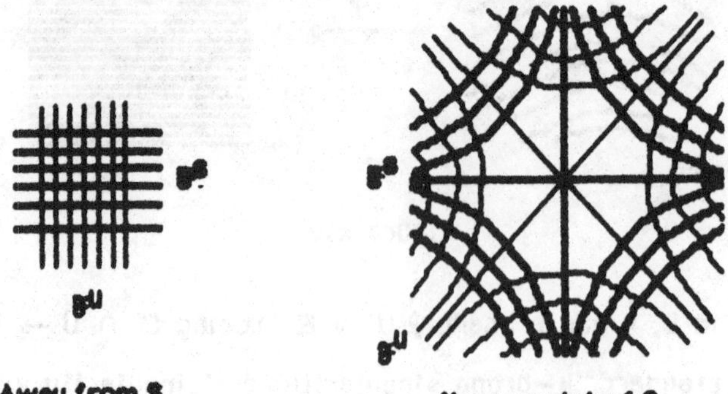

Figure 6.4

Lemma 6.2: Let $h:F \to F$ be a non-periodic irreducible automorphism of a closed oriented hyperbolic surface. Then h is isotopic to an automorphism $h_*:F \to F$ such that $h_*(\mathcal{F}^s) = \mathcal{F}^s$ and $h_*(\mathcal{F}^u) = \mathcal{F}^u$ for some pair of transverse singular foliations $\mathcal{F}^s, \mathcal{F}^u$.

Remark: As before, h_* is not necessarily a diffeomorphism.

Proof: Define an equivalence relation ~ on F, where $x \sim y$ if either

i) x, y are in the closure of the same component of $F - (L^s \cup L^u)$.

or ii) x, y are in the closure of the same component of $L^s - L^u$.

or iii) x, y are in the closure of the same component of $L^u - L^s$.

or iv) $x = y$.

We shall show that $F/\sim$ is a closed surface homeomorphic to F, and that the laminations L^s, L^u project to transverse singular foliations $\mathcal{F}^s, \mathcal{F}^u$.

Let π denote the projection from F to $F/\sim$. For any point x in F, we describe a small neighborhood of $\pi(x)$ in $F/\sim$. First suppose that x is on non-boundary leaves γ of L^s and δ of L^u. Let σ and τ be short closed segments of γ and δ respectively centered on x. Associated with the Cantor sets $\sigma \cap L^u$ and $\tau \cap L^s$ are Cantor functions $\alpha:\sigma \to [-1,1]$ and $\beta:\tau \to [-1,1]$, chosen so that $\alpha(x) = \beta(x) = 0$. If ϵ is sufficiently small, α and β induce a homeomorphism φ from a rectangle neighborhood U of $\pi(x)$ onto $[-\epsilon,\epsilon] \times [-\epsilon,\epsilon]$. The composition $\varphi \cdot \pi$ takes the components of $U \cap L^s$ to horizontal intervals and the components of $U \cap L^u$ to vertical intervals (see Figure 6.5).

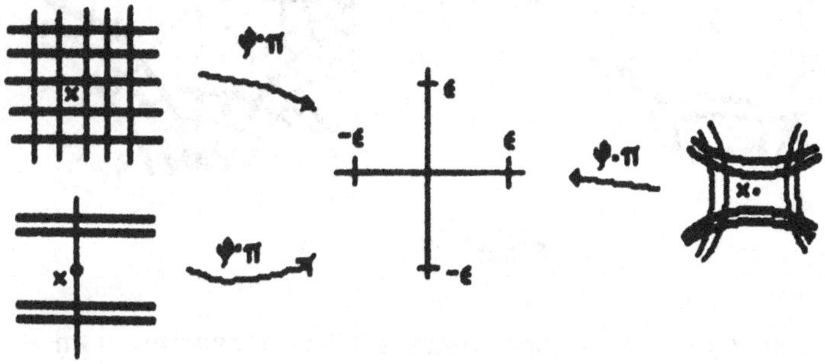

Figure 6.5

For x in a component of $L^s - L^u$ or $L^u - L^s$, charts to $[-\epsilon,\epsilon] \times [-\epsilon,\epsilon]$ are formed from two "half-charts" as follows. For x on a leaf δ in $L^u - L^s$, say, a "half-chart" to $[-\epsilon, \epsilon] \times [0,\epsilon]$ (or $[-\epsilon,\epsilon] \times [-\epsilon,0]$) is formed as above starting from the intersection of δ with the boundary leaf of L^s nearest x. Similarly, rectangle components of $F-(L^s \cup L^u)$ give rise to charts formed from "quarter-charts", and so on for 2k-gons (see Figure 6.6).

Note that the singular points arise from the 2k-gons, as the associated charts give the standard V_k model.

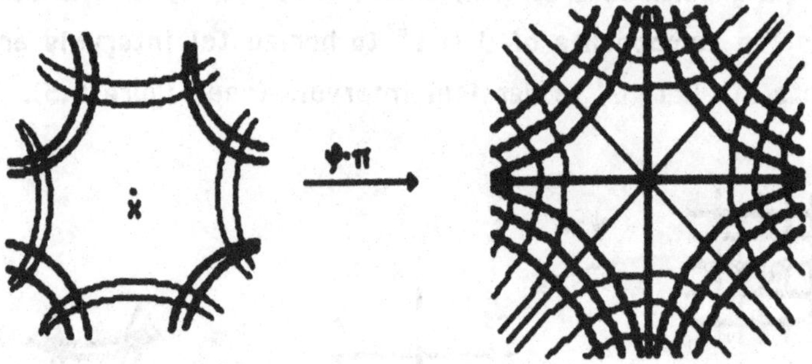

Figure 6.6

Exercise: Show that π is approximable by a homeomorphism Θ from $F/\sim$ onto F.

The laminations L^s, L^u project to transverse singular foliations on $F/\sim$; let $\mathcal{F}^s$ and $\mathcal{F}^u$ be $\Theta^{-1}\cdot\pi(L^s)$ and $\Theta^{-1}\cdot\pi(L^u)$ respectively. By

Theorem 6.1, h is isotopic to $h':F \to F$ preserving L^s, L^u. It follows that h' induces a homeomorphism h'' of $F/\sim$ preserving $\pi(L^s)$ and $\pi(L^u)$. Then h_*, the required homeomorphism, is given by $\Theta^{-1} \cdot h'' \cdot \Theta : F \to F$.

Definition: A *transverse measure* μ to a singular foliation $\mathcal{F}$ defines on each arc α transverse to $\mathcal{F}$ a non-negative Borel measure $\mu|\alpha$ with the following properties.

1) If β is a subarc of α, then $\mu|\beta$ is the restriction of $\mu|\alpha$.

2) If α_0, α_1 are arcs transverse to $\mathcal{F}$ related by a homotopy $\alpha: I \times I \longrightarrow F$ such that $\alpha(I \times 0) = \alpha_0$, $\alpha(I \times 1) = \alpha_1$ and $\alpha(a \times I)$ is contained in a leaf of $\mathcal{F}$ for all $a \in I$, then $\mu|\alpha_0 = \mu|\alpha_1$.

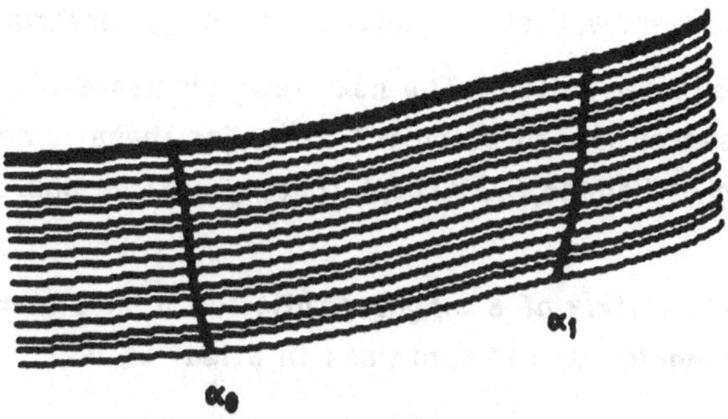

Figure 6.7

Definition: An automorphism h of a closed orientable surface is *pseudo-Anosov* if there are transverse singular foliations $\mathcal{F}^s$, $\mathcal{F}^u$ equipped with transverse measures μ^s, μ^u such that

$$h(\mathcal{F}^s,\mu^s) = (\mathcal{F}^s, \lambda\mu^s),$$
$$h(\mathcal{F}^u,\mu^u) = (\mathcal{F}^u,\lambda^{-1}\mu^u)$$

for some $\lambda > 1$.

Remark: Here $h(\mu)$ denotes the transverse measure defined by the equation $h(\mu)|h(\alpha) = \mu|\alpha$. If instead we had defined $h(\mu)$ by $h(\mu)|\alpha = \mu|h(\alpha)$ then λ would be replaced by λ^{-1} in the above definition. Our definiton has the consequence that $(h_1 \cdot h_2)(\mu) = h_1(h_2(\mu))$ rather than $h_2(h_1(\mu))$.

Theorem 6.3: Every non-periodic irreducible automorphism of a closed orientable hyperbolic surface is isotopic to a pseudo-Anosov automorphism.

Remark: We shall show that the automorphism h_* constructed in Lemma 6.2 is pseudo-Anosov. The next few lemmas are used to construct the transverse measures on $\mathcal{F}^s$, $\mathcal{F}^u$. For these lemmas we shall use the notation of Lemma 6.2 except that h_* will be abbreviated to h.

Definition: A *separatrix* of a singular foliation $\mathcal{F}$ is a maximal arc beginning at a singularity and contained in a leaf of $\mathcal{F}$.

Lemma 6.4: The foliations $\mathcal{F}^s$ and $\mathcal{F}^u$ have no closed leaves; moreover each separatrix has only one singular point and is dense in the surface. If α is a separatrix of $\mathcal{F}^u$ beginning at a singularity s, then there is an integer $m > 0$ such that $h^m(\sigma) \subset \sigma$ and for all $x \in \sigma$, $h^m(x)$ is in the open subinterval (s,x) of σ. A similar result holds for the separatrices of $\mathcal{F}^s$ (with h^{-1} replacing h).

Proof: This follows directly from the construction: details are left to the reader.

Lemma 6.5: There are closed sets A, B $\subset$ F with the following properties.

1) A is a union of closed arcs each contained in a separatrix of $\mathcal{F}^s$.

2) B is a union of closed arcs each contained in a separatrix of $\mathcal{F}^u$.

3) Each component of F − (A ∪ B) is homeomorphic to (0,1)×(0,1) by a homeomorphism carring leaves of $\mathcal{F}^s$ and $\mathcal{F}^u$ to horizontals and verticals respectively (compare Figure 6.4).

4) $h^{-1}(A) \subset A$.

5) $h(B) \subset B$.

Proof: For each separatrix σ of $\mathcal{F}^u$, $B \cap \sigma$ will be a closed arc β_σ beginning at the singularity of σ. It follows from Lemma 6.4 that the β_σ's may be chosen so that $h(\beta_\sigma) \subset \beta_{h(\sigma)}$. Observe that conditions 2) and 5) above are satisfied.

Let A be the set of points in F which can be joined to an endpoint of some β_σ by an arc of a leaf of $\mathcal{F}^s$ in F−B. Let U be a component of F − (A ∪ B). Any point $x \in U$ is in the interior of a closed segment α_x of a leaf of $\mathcal{F}^s$ whose endpoints lie in the interiors of the (not necessarily distinct) arcs β_{σ_1} and β_{σ_2}. It follows from the construction of A that β_{σ_1} and β_{σ_2} do not depend on the choice of $x \in U$.

The reader may now check that condition 3) above is satisfied.

We want every component of A to contain a singularity to ensure that conditions 1) and 4) are satisfied. For this we must modify the sets A and B somewhat. If some component of A contains no singularity then it consists of an closed arc α whose endpoints are interior points of β_{σ_1} and β_{σ_2}. Such a component necessarily contains an endpoint of some $\beta_\sigma \subset B$. We shall modify A and B by shrinking β_σ towards its singularity and sliding α to a parallel arc α' (see Figure 6.8).

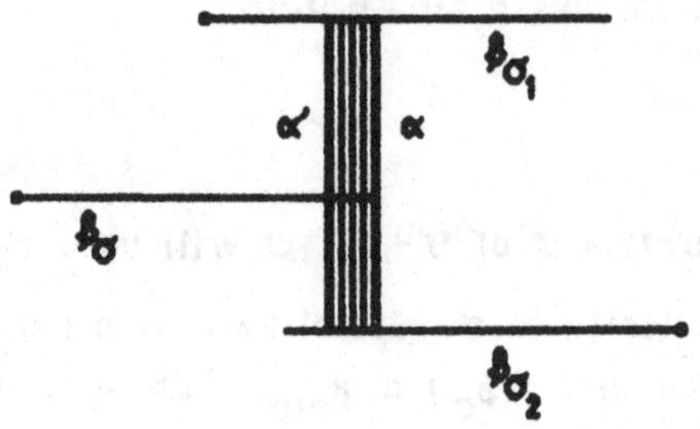

Figure 6.8

We continue shrinking β_σ until α' becomes amalgamated with another component of A. Since this operation reduces the number of components of A, a finite number of iterations yields a set A with a singularity in each component. It follows that conditions 1) and 4) are now satisfied.

Definition: A *rectangle* R is a map $\rho: I \times I \to F$ such that ρ is an embedding on the interior, $\rho(\text{point} \times I)$ is contained in a leaf of $\mathcal{F}^u$, and $\rho(I \times \text{point})$ is contained in a leaf of $\mathcal{F}^s$. For a rectangle R denote $\rho(\partial I \times I)$ by $\partial^u R$ and $\rho(I \times \partial I)$ by $\partial^s R$. We often use the same notation for a rectangle and its image in F.

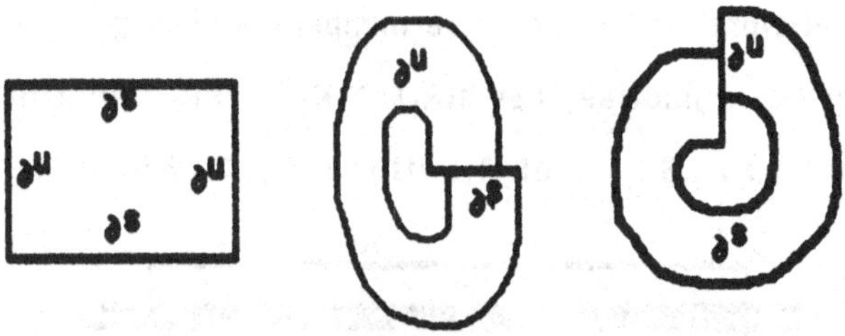

Rectangles

Figure 6.9

Corollary 6.5.1: There is a decomposition of F into a finite union of rectangles $R_1, \ldots, R_n$ with the following properties.

1) If $i \neq j$, Int $R_i \cap$ Int R_j is empty.

2) $h\left(\bigcup_{j=1}^{n} \partial^u R_j\right) \subset \bigcup_{j=1}^{n} \partial^u R_j$.

3) $h^{-1}\left(\bigcup_{i=1}^{n} \partial^s R_i\right) \subset \bigcup_{i=1}^{n} \partial^s R_i$.

Remark: Such a decomposition is called a *Markov partition* for h.

Proof of 6.3: Recall that our aim is to construct transverse measures μ^s, μ^u for $\mathcal{F}^s$, $\mathcal{F}^u$ such that $h(\mu^s) = \lambda\mu^s$ and $h(\mu^u) = \lambda^{-1}\mu^u$ for some $\lambda > 1$. These measures will assign to each rectangle R_i a "height" y_i and a "width" x_i; for example, y_i will be the μ^s measure of any "vertical" cross-section of R_i. The first step is to determine necessary conditions on the positive numbers x_i and y_i.

Corollary 6.5.1 implies that $h(R_i) \cap R_j$ consists of finitely many subrectangles $S_1, \ldots, S_{\ell(i,j)}$ of R_j with $\partial^u S_k \subset \partial^u R_j$ (see Fig. 6.10).

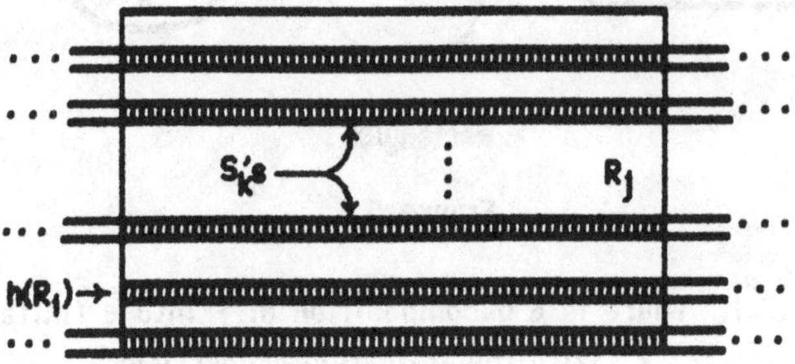

$h(R_i) \cap R_j$

Figure 6.10

Let $a_{ij} = \ell(i,j)$, the number of subrectangles making up $h(R_i) \cap R_j$, and let A be the square matrix (a_{ij}). By definition, each vertical cross-section of $h(R_i)$ has $h(\mu^s)$ measure equal to y_i. Since we are requiring that $h(\mu^s) = \lambda\mu^s$ each S_k has height $\lambda^{-1}y_i$ (measured by μ^s).

Since R_j is the union of the sets $h(R_i) \cap R_j$ $(1 \leq i \leq n)$,

$$y_j = \sum_{i=1}^{n} a_{ij}\lambda^{-1}y_i.$$

Therefore the column vector $y = (y_i)$ is an eigenvector of A with eigenvalue λ. Similarly, the column vector $x = (x_i)$ is an eigenvector of A^t with eigenvalue λ^{-1}. Next we show that these necessary conditions can be satisfied.

Lemma 6.6: The matrix A has an eigenvector y with $y_1, \ldots, y_n > 0$, corresponding to an eigenvalue $\lambda > 1$.

Proof: Let C be the cone consisting of those vectors $z \in \mathbb{R}^n$ with non-negative coordinates. Since the transformation $z \rightarrow Az$ leaves C invariant, the Brouwer fixed point theorem implies that C contains an eigenvector y of A.

Suppose some coordinate of y, say y_i, is zero. Since $y \neq 0$, it has a non-zero coordinate, say y_j. It follows that the (i,j) entry in any positive power of A is zero, implying that $h^m(R_i)$ is disjoint from R_j for all $m > 0$. But the union of the sets $h^m(R_i)$ $(m = 1, 2, 3, \ldots)$ contains a separatrix of $\mathcal{F}^s$, which is dense in F by Lemma 6.4. This contradiction shows that each entry of y is strictly positive.

The argument above shows that for some $m > 0$, all the entries of

A^m are strictly positive integers. It follows that the eigenvalue λ corresponding to y is strictly greater than 1.

Returning to the proof of Theorem 6.3, let α be a closed arc transverse to $\mathcal{F}^s$. For each $m > 0$, α is the union of the sets $\alpha \cap h^m(R_i)$; let $u_{i,m}$ denote the number of components of $\alpha \cap h^m(R_i)$.

$$\text{Define}\quad \mu^s(\alpha) \;=\; \lim_{m\to\infty} \sum_{i=1}^{n} \lambda^{-m} y_i u_{i,m}.$$

This limit exists because the "error" in the sum arises entirely from the components of $\alpha \cap h^m(R_i)$ containing the endpoints of α. The reader may verify that this does define a measure μ^s transverse to $\mathcal{F}^s$ such that $h(\mu^s) = \lambda\mu^s$.

The same construction applied to h^{-1} yields a measure μ^u transverse to $\mathcal{F}^u$ such that $h^{-1}(\mu^u) = \bar{\lambda}\,\mu^u$, for some $\bar{\lambda} > 1$. It follows that $h(\mu^u) = \bar{\lambda}^{-1}\mu^u$. Let μ be the product measure $\mu^s \times \mu^u$ on F; observe that $\mu(F) = x^t y$ is finite and non-zero. Therefore

$$\begin{aligned}
\mu(F) &= [h(\mu)](F) \\
&= [h(\mu^s) \times h(\mu^u)](F) \\
&= [\lambda\mu^s \times \bar{\lambda}^{-1}\mu^u](F) \\
&= \lambda\bar{\lambda}^{-1}[\mu^s \times \mu^u](F), \\
&= \lambda\bar{\lambda}^{-1}\mu(F)
\end{aligned}$$

implying that $\bar{\lambda} = \lambda$.

References

[CL] A. Casson, D. Long, Algorithmic compression of surface automorphisms, *Invent. math.* 81 (1985), 295–303.

[FLP] A. Fathi, F. Laudenbach, V. Poenaru, Travaux de Thurston sur les surfaces, Asterisque 66–67 (1979).

[F_1] W. Fenchel, Estensioni gruppi discontinui e transformazioni periodiche delle surficie, *Rend. Acc. Naz. Lincei (se.fis., mat. e. nat)* 5 (1948), 326–329.

[F_2] W. Fenchel, Bemarkongen om endelige gruppen af abbilungs-klasser, *Mat. Tidskrift* B 1950, 90–95.

[HT] M. Handel, W. Thurston, New proofs of some results of Nielsen, *Adv. Math.* 56 (1985), 173–191.

[K] S. Kerckhoff, The Nielsen realization problem, *Annals of Math.* 117 (1983), 235–265.

[L] D. Long, *Cobordism of knots and surface automorphisms,* thesis, Cambridge 1983.

[M] S. Morita, Characteristic classes of surface bundles, *Bull. AMS* 11, No. 2 (1984), 386–388

[N_1] J. Nielsen, Untersuchungen zur Topologie der geschlossenen zweiseitigen Flachen I, *Acta Math.* 50 (1927), 189–358.

[N_2] J. Nielsen, Untersuchungen zur Topologie der geschlossenen zweiseitigen Flachen II, *Acta Math.* 53 (1929), 1–76.

[N_3] J. Nielsen, Untersuchungen zur Topologie der geschlossenen zweiseitigen Flachen III, *Acta Math.* 58 (1931), 87–167.

[N_4] J. Nielsen, Abbildungsklassen endlicher Ordnung, *Acta Math.* (1942), 23–115.

[P] R. Penner, *A computation of the action of the mapping class groups on isotopy classes of curves and arcs in surfaces*, thesis, M.I.T. 1982

[T] W. Thurston, *On the geometry and dynamics of diffeomorphisms of surfaces I*, preprint.

Index